T0235701

Israel Kleiner

A History of
Abstract Algebra

Birkhäuser
Boston • Basel • Berlin

Israel Kleiner
Department of Mathematics and Statistics
York University
Toronto, ON M3J 1P3
Canada
kleiner@rogers.com

Cover design by Alex Gerasev, Revere, MA.

Mathematics Subject Classification (2000): 00-01, 00-02, 01-01, 01-02, 01A55, 01A60, 01A70, 12-03, 13-03, 15-03, 16-03, 20-03, 97-03

Library of Congress Control Number: 2007932362

ISBN-13: 978-0-8176-4684-4   e-ISBN-13: 978-0-8176-4685-1

Printed on acid-free paper.

9 8 7 6 5 4 3 2 1

www.birkhauser.com   (LAP/EB)

With much love to my family
Nava
Ronen, Melissa, Leeor, Tania, Ayelet, Tamir
Tia, Jordana, Jake

Contents

Preface

My goal in writing this book was to give an account of the history of many of the basic concepts, results, and theories of abstract algebra, an account that would be useful for teachers of relevant courses, for their students, and for the broader mathematical public.

The core of a first course in abstract algebra deals with groups, rings, and fields. These are the contents of Chapters 2, 3, and 4, respectively. But abstract algebra grew out of an earlier classical tradition, which merits an introductory chapter in its own right (Chapter 1). In this tradition, which developed before the nineteenth century, "algebra" meant the study of the solution of polynomial equations. In the twentieth century it meant the study of abstract, axiomatic systems such as groups, rings, and fields. The transition from "classical" to "modern" occurred in the nineteenth century. Abstract algebra came into existence largely because mathematicians were unable to solve classical (pre-nineteenth-century) problems by classical means. The classical problems came from number theory, geometry, analysis, the solvability of polynomial equations, and the investigation of properties of various number systems. A major theme of this book is to show how "abstract" algebra has arisen in attempts to solve some of these "concrete" problems, thus providing confirmation of Whitehead's paradoxical dictum that "the utmost abstractions are the true weapons with which to control our thought of concrete fact." Put another way: there is nothing so practical as a good theory.

Although linear algebra is not normally taught in a course in abstract algebra, its evolution has often been connected with that of groups, rings, and fields. And, of course, vector spaces are among the fundamental notions of abstract algebra. This warrants a (short) chapter on the history of linear algebra (Chapter 5).

Abstract algebra is essentially a creation of the nineteenth century, but it became an independent and flourishing subject only in the early decades of the twentieth, largely through the pioneering work of Emmy Noether, who has been called "the father" of abstract algebra. Thus the chapter on Noether's algebraic work (Chapter 6).

It is my firm belief, buttressed by my own teaching experience, that the history of mathematics can make an important contribution to our—teachers' and students'—understanding and appreciation of mathematics. It can act as a useful integrating component in the teaching of any area of mathematics, and can provide motivation and perspective. History points to the sources of the subject, hence to some of its central notions. It considers the context in which the originator of an idea was working in order to bring to the fore the "burning problem" which he or she was trying to solve.

The biologist Ernest Haeckel's fundamental principle that "ontogeny recapitulates phylogeny"—that the development of an individual retraces the evolution of its species—was adapted by George Polya, as follows: "Having understood how the human race has acquired the knowledge of certain facts or concepts, we are in a better position to judge how [students] should acquire such knowledge." This statement is but one version of the so-called "genetic principle" in mathematics education. As Polya notes, one should view it as a guide to, not a substitute for, judgment. Indeed, it is the teacher who knows best when and how to use historical material in the classroom, if at all. Chapter 7 describes a course in abstract algebra inspired by history. I have taught it in an in-service Master's Program for high school teachers of mathematics, but it can be adapted to other types of algebra courses.

In each of the above chapters I mention the major contributors to the development of algebra. To emphasize the human face of the subject, I have included a chapter on the lives and works of six of its major creators: Cayley, Dedekind, Galois, Gauss, Hamilton, and Noether (Chapter 8). This is a substantial chapter—in fact, the longest in the book. Each of the biographies is a mini-essay, since I wanted to go beyond a mere listing of names, dates, and accomplishments.

The concepts of abstract algebra did not evolve independently of one another. For example, field theory and commutative ring theory have common sources, as do group theory and field theory. I wanted, however, to make the chapters independent, so that a reader interested in finding out about, say, the evolution of field theory would not need to read the chapter on the evolution of ring theory. This has resulted in a certain amount of repetition in some of the chapters.

The book is not meant to be a primer of abstract algebra from which students would learn the elements of groups, rings, or fields. Neither abstract algebra nor its history are easy subjects. Most students will probably need the guidance of a teacher on a first reading.

To enhance the usefulness of the book, I have included many references, for the most part historical. For ease of use, they are placed at the end of each chapter. The historical references are mainly to secondary sources, since these are most easily accessible to teachers and students. Many of these secondary sources contain references to primary sources.

The book is a far-from-exhaustive account of the history of abstract algebra. For example, while I devote a mere twenty pages or so to the history of groups, an entire book has been published on the topic. My main aim was to give an overview of many of the basic ideas of abstract algebra taught in a first course in the subject. For readers who want to pursue the subject further, I have indicated in the body of each chapter where additional material can be found. Detection of errors in the historical account will be gratefully acknowledged.

The primary audience for the book, as I see it, is teachers of courses in abstract algebra. I have noted some of the uses they may put it to. The book can also be used in courses on the history of mathematics. And it may appeal to algebraists who want to familiarize themselves with the history of their subject, as well as to the broader mathematical community.

Finally, I want to thank Ann Kostant, Elizabeth Loew, and Avanti Paranjpye of Birkhäuser for their outstanding cooperation in seeing this book to completion.

Israel Kleiner
Toronto, Ontario
May 2007

Permissions

Grateful acknowledgment is hereby given for permission to reprint in full or in part, with minor changes, the following:

I. Kleiner, "Algebra." *History of Modern Science and Mathematics*, Scribner's, 2002, pp. 149–167. Reprinted with permission of Thomson Learning: www.thomsonrights.com. (Used in Chapters 1 and 5.)

I. Kleiner, "The evolution of group theory: a brief survey." *Mathematics Magazine* **6** (1986) 195–215. Reprinted with permission of the Mathematical Association of America. (Used in Chapter 2.)

I. Kleiner, "From numbers to rings: the early history of ring theory." *Elemente der Mathematik* **53** (1998) 18–35. Reprinted with permission of Birkhäuser. (Used in Chapter 3.)

I. Kleiner, "Field theory: from equations to axiomatization," Parts I and II. *American Mathematical Monthly* **106** (1999) 677–684 and 859–863. Reprinted with permission of the Mathematical Association of America. (Used in Chapter 4.)

I. Kleiner, "Emmy Noether: highlights of her life and work." *L'Enseignement Mathématique* **38** (1992) 103–124. (Used in Chapters 6 and 8.)

I. Kleiner, "A historically focused course in abstract algebra." *Mathematics Magazine* **71** (1998) 105–111. Reprinted with permission of the Mathematical Association of America. (Used in Chapter 7.)

1

History of Classical Algebra

1.1 Early roots

For about three millennia, until the early nineteenth century, "algebra" meant solving polynomial equations, mainly of degree four or less. Questions of notation for such equations, the nature of their roots, and the laws governing the various number systems to which the roots belonged, were also of concern in this connection. All these matters became known as *classical algebra*. (The *term* "algebra" was only coined in the ninth century **AD**.) By the early decades of the twentieth century, algebra had evolved into the study of axiomatic systems. The axiomatic approach soon came to be called *modern* or *abstract algebra*. The transition from classical to modern algebra occurred in the nineteenth century.

Most of the major ancient civilizations, the Babylonian, Egyptian, Chinese, and Hindu, dealt with the solution of polynomial equations, mainly linear and quadratic equations. The Babylonians (c. 1700 **BC**) were particularly proficient "algebraists." They were able to solve quadratic equations, as well as equations that lead to quadratic equations, for example $x + y = a$ and $x^2 + y^2 = b$, by methods similar to ours. The equations were given in the form of "word problems." Here is a typical example and its solution:

> I have added the area and two-thirds of the side of my square and it is 0;35 [35/60 in sexagesimal notation]. What is the side of my square?

In modern notation the problem is to solve the equation $x^2 + (2/3)x = 35/60$. The solution given by the Babylonians is:

> You take 1, the coefficient. Two-thirds of 1 is 0;40. Half of this, 0;20, you multiply by 0;20 and it [the result] 0;6,40 you add to 0;35 and [the result] 0;41,40 has 0;50 as its square root. The 0;20, which you have multiplied by itself, you subtract from 0;50, and 0;30 is [the side of] the square.

The instructions for finding the solution can be expressed in modern notation as $x = \sqrt{[(0;40)/2]^2 + 0;35} - (0;40)/2 = \sqrt{0;6,40 + 0;35} - 0;20 = \sqrt{0;41,40} - 0;20 = 0;50 - 0;20 = 0;30$.

These instructions amount to the use of the formula $x = \sqrt{(a/2)^2 + b} - a/2$ to solve the equation $x^2 + ax = b$. This is a remarkable feat. See [1], [8].

The following points about Babylonian algebra are important to note:

(a) There was no algebraic notation. All problems and solutions were verbal.
(b) The problems led to equations with numerical coefficients. In particular, there was no such thing as a general quadratic equation, $ax^2 + bx + c = 0$, with a, b, and c arbitrary parameters.
(c) The solutions were *prescriptive*: do such and such and you will arrive at the answer. Thus there was no *justification* of the procedures. But the accumulation of example after example of the same type of problem indicates the existence of some form of justification of Babylonian mathematical procedures.
(d) The problems were chosen to yield only positive rational numbers as solutions. Moreover, only one root was given as a solution of a quadratic equation. Zero, negative numbers, and irrational numbers were not, as far as we know, part of the Babylonian number system.
(e) The problems were often phrased in geometric language, but they were not problems *in* geometry. Nor were they of practical use; they were likely intended for the training of students. Note, for example, the addition of the area to 2/3 of the side of a square in the above problem. See [2], [6], [14], [18] for aspects of Babylonian algebra.

The Chinese (c. 200 BC) and the Indians (c. 600 BC) advanced beyond the Babylonians (the dates for both China and India are very rough). For example, they allowed negative coefficients in their equations (though not negative roots), and admitted two roots for a quadratic equation. They also described procedures for manipulating equations, but had no notation for, nor justification of, their solutions. The Chinese had methods for approximating roots of polynomial equations of any degree, and solved systems of linear equations using "matrices" (rectangular arrays of numbers) well before such techniques were known in Western Europe. See [7], [10], [18].

1.2 The Greeks

The mathematics of the ancient Greeks, in particular their geometry and number theory, was relatively advanced and sophisticated, but their algebra was weak. Euclid's great work *Elements* (c. 300 BC) contains several parts that have been interpreted by historians, *with notable exceptions* (e.g., [14, 16]), as algebraic. These are geometric propositions that, if translated into algebraic language, yield algebraic results: laws of algebra as well as solutions of quadratic equations. This work is known as *geometric algebra*.

For example, Proposition II.4 in the *Elements* states that "If a straight line be cut at random, the square on the whole is equal to the square on the two parts and twice the rectangle contained by the parts." If a and b denote the parts into which the straight line is cut, the proposition can be stated algebraically as $(a + b)^2 = a^2 + 2ab + b^2$.

Proposition II.11 states: "To cut a given straight line so that the rectangle contained by the whole and one of the segments is equal to the square on the remaining segment." It asks, in algebraic language, to solve the equation $a(a - x) = x^2$. See [7, p. 70].

Note that Greek algebra, such as it is, speaks of *quantities* rather than numbers. Moreover, *homogeneity* in algebraic expressions is a strict requirement; that is, all terms in such expressions must be of the same degree. For example, $x^2 + x = b^2$ would not be admitted as a legitimate equation. See [1], [2], [18], [19].

A much more significant Greek algebraic work is Diophantus' *Arithmetica* (c. 250 AD). Although essentially a book on number theory, it contains solutions of equations in integers or rational numbers. More importantly for progress in algebra, it introduced a partial algebraic notation—a most important achievement: ς denoted an unknown, Φ negation, ἴσ equality, Δ^σ the square of the unknown, K^σ its cube, and M the absence of the unknown (what we would write as x^0). For example, $x^3 - 2x^2 + 10x - 1 = 5$ would be written as $K^\sigma \alpha \varsigma \text{í} \Phi \Delta^\sigma \beta M \alpha \text{íσ} M \varepsilon$ (numbers were denoted by letters, so that, for example, α stood for 1 and ε for 5; moreover, there was no notation for addition, thus all terms with positive coefficients were written first, followed by those with negative coefficients).

Diophantus made other remarkable advances in algebra, namely:

(a) He gave two basic rules for working with algebraic expressions: the transfer of a term from one side of an equation to the other, and the elimination of like terms from the two sides of an equation.

(b) He defined negative powers of an unknown and enunciated the law of exponents, $x^m x^n = x^{m+n}$, for $-6 \le m, n, m + n \le 6$.

(c) He stated several rules for operating with negative coefficients, for example: "deficiency multiplied by deficiency yields availability" $((-a)(-b) = ab)$.

(d) He did away with such staples of the classical Greek tradition as (i) giving a geometric interpretation of algebraic expressions, (ii) restricting the product of terms to degree at most three, and (iii) requiring homogeneity in the terms of an algebraic expression. See [1], [7], [18].

1.3 Al-Khwarizmi

Islamic mathematicians attained important algebraic accomplishments between the ninth and fifteenth centuries AD. Perhaps foremost among them was Muhammad ibn-Musa al-Khwarizmi (c. 780–850), dubbed by some "the Euclid of algebra" because he systematized the subject (as it then existed) and made it into an independent field of study. He did this in his book *al-jabr w al-muqabalah*. "Al-jabr" (from which stems our word "algebra") denotes the moving of a negative term of an equation to the other side so as to make it positive, and "al-muqabalah" refers to cancelling equal (positive) terms on both sides of an equation. These are, of course, basic procedures for solving polynomial equations. Al-Khwarizmi (from whose name the term "algorithm" is derived) applied them to the solution of quadratic equations. He classified these into five types: $ax^2 = bx$, $ax^2 = b$, $ax^2 + bx = c$, $ax^2 + c = bx$, and $ax^2 = bx + c$. This categorization was necessary since al-Khwarizmi did not admit negative coefficients or zero. He also had essentially no notation, so that his problems and solutions were expressed rhetorically. For example, the first and third equations above were given as: "squares equal roots" and "squares and roots equal numbers" (an unknown was called a "root"). Al-Khwarizmi did offer justification, albeit geometric, for his solution procedures. See [13], [17].

Muhammad al-Khwarizmi (ca 780–850)

The following is an example of one of his problems with its solution. [7, p. 245]: "What must be the square, which when increased by ten of its roots amounts to thirty-nine?" (i.e., solve $x^2 + 10x = 39$).

Solution: "You halve the number of roots [the coefficient of x], which in the present instance yields five. This you multiply by itself; the product is twenty-five. Add this to thirty nine; the sum is sixty-four. Now take the root of this, which is eight, and subtract from it half the number of the roots, which is five; the remainder is three. This is the root of the square which you sought." (Symbolically, the prescription is: $\sqrt{[(1/2) \times 10]^2 + 39} - (1/2) \times 10$.)

Here is al-Khwarizmi's justification: Construct the gnomon as in Fig. 1, and "complete" it to the square in Fig. 2 by the addition of the square of side 5. The resulting square has length $x + 5$. But it also has length 8, since $x^2 + 10x + 5^2 = 39 + 25 = 64$. Hence $x = 3$.

Now a brief word about some contributions of mathematicians of Western Europe of the fifteenth and sixteenth centuries. Known as "abacists" (from "abacus") or "cossists" (from "cosa," meaning "thing" in Latin, used for the unknown), they extended,

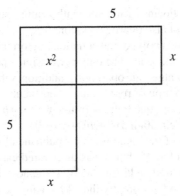

Fig. 1.

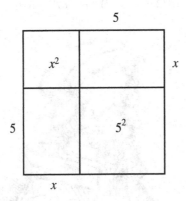

Fig. 2.

and generally improved, previous notations and rules of operation. An influential work of this kind was Luca Pacioli's *Summa* of 1494, one of the first mathematics books in print (the printing press was invented in about 1445). For example, he used "*co*" (cosa) for the unknown, introducing symbols for the first 29 (!) of its powers, "*p*" (piu) for plus and "*m*" (meno) for minus. Others used $\mathbf{R}_x$ (radix) for square root and $\mathbf{R}_{x.3}$ for cube root. In 1557 Robert Recorde introduced the symbol "=" for equality with the justification that "noe 2 thynges can be moare equalle." See [7], [13], [17].

1.4 Cubic and quartic equations

The Babylonians were solving quadratic equations by about 1600 BC, using essentially the equivalent of the quadratic formula. A natural question is therefore whether cubic equations could be solved using similar formulas (see below). Another three thousand years would pass before the answer would be known. It was a great event in algebra when mathematicians of the sixteenth century succeeded in solving "by radicals" not only cubic but also quartic equations.

A solution by radicals of a polynomial equation is a formula giving the roots of the equation in terms of its coefficients. The only permissible operations to be applied to the coefficients are the four algebraic operations (addition, subtraction, multiplication, and division) and the extraction of roots (square roots, cube roots, and so on, that is, "radicals"). For example, the quadratic formula $x = (-b \pm \sqrt{b^2 - 4ac})/2a$ is a solution by radicals of the equation $ax^2 + bx + c = 0$.

A solution by radicals of the cubic was first published by Cardano in *The Great Art* (referring to algebra) of 1545, but it was discovered earlier by del Ferro and by Tartaglia. The latter had passed on his method to Cardano, who had promised that he would not publish it, which he promptly did. What came to be known as *Cardano's formula* for the solution of the cubic $x^3 = ax + b$ was given by

$$x = \sqrt[3]{b/2 + \sqrt{(b/2)^2 - (a/3)^3}} + \sqrt[3]{b/2 - \sqrt{(b/2)^2 - (a/3)^3}}.$$

Girolamo Cardano (1501–1576)

Several comments are in order:

(i) Cardano used no symbols, so his "formula" was given rhetorically (and took up close to half a page). Moreover, the equations he solved all had *numerical* coefficients.

(ii) He was usually satisfied with finding a single root of a cubic. In fact, if a proper choice is made of the cube roots involved, then all three roots of the cubic can be determined from his formula.

(iii) Negative numbers are found occasionally in his work, but he mistrusted them, calling them "fictitious." The coefficients and roots of the cubics he considered were *positive* numbers (but he admitted irrationals), so that he viewed (say) $x^3 = ax + b$ and $x^3 + ax = b$ as distinct, and devoted a chapter to the solution of each (compare al-Khwarizmi's classification of quadratics).

(iv) He gave *geometric* justifications of his solution procedures for the cubic.

The solution by radicals of polynomial equations of the fourth degree (quartics) soon followed. The key idea was to reduce the solution of the quartic to that of a cubic. Ferrari was the first to solve such equations, and his work was included in Cardano's *The Great Art*. See [1], [7], [10], [12]

It should be pointed out that methods for finding *approximate* roots of cubic and quartic equations were known well before such equations were solved by radicals. The latter solutions, though exact, were of little practical value. However, the ramifications of these "impractical" ideas of mathematicians of the Italian Renaissance were very significant, and will be considered in Chapter 2.

1.5 The cubic and complex numbers

Mathematicians adhered for centuries to the following view with respect to the square roots of negative numbers: since the squares of positive as well as of negative numbers are positive, square roots of negative numbers do not—in fact, cannot—exist. All this changed following the solution by radicals of the cubic in the sixteenth century.

Square roots of negative numbers arise "naturally" when Cardano's formula (see p. 6) is used to solve cubic equations. For example, application of his formula to the equation $x^3 = 9x + 2$ gives $x = \sqrt[3]{2/2 + \sqrt{(2/2)^2 - (9/3)^3}} + \sqrt[3]{2/2 - \sqrt{(2/2)^2 - (9/3)^3}} = \sqrt[3]{1 + \sqrt{-26}} + \sqrt[3]{1 - \sqrt{-26}}$. What is one to make of this solution? Since Cardano was suspicious of negative numbers, he certainly had no taste for their square roots, so he regarded his formula as inapplicable to equations such as $x^3 = 9x + 2$. Judged by past experience, this was not an unreasonable attitude. For example, to the Pythagoreans, the side of a square of area 2 was nonexistent (in today's language, we would say that the equation $x^2 = 2$ is unsolvable).

All this was changed by Bombelli. In his important book *Algebra* (1572) he applied Cardano's formula to the equation $x^3 = 15x + 4$ and obtained $x = \sqrt[3]{2 + \sqrt{-121}} + \sqrt[3]{2 - \sqrt{-121}}$. But he could not dismiss the solution, for he noted by inspection that $x = 4$ is a root of this equation. Moreover, its other two roots, $-2 \pm \sqrt{3}$, are also real numbers. Here was a paradox: while all three roots of the cubic $x^3 = 15x + 4$ are real, the formula used to obtain them involved square roots of negative numbers—meaningless at the time. How was one to resolve the paradox?

Bombelli adopted the rules for real quantities to manipulate "meaningless" expressions of the form $a + \sqrt{-b}$ $(b > 0)$ and thus managed to show that $\sqrt[3]{2 + \sqrt{-121}} = 2 + \sqrt{-1}$ and $\sqrt[3]{2 - \sqrt{-121}} = 2 - \sqrt{-1}$, and hence that $x = \sqrt[3]{2 + \sqrt{-121}} + \sqrt[3]{2 - \sqrt{-121}} = (2 + \sqrt{-1}) + (2 - \sqrt{-1}) = 4$. Bombelli had given meaning to the "meaningless" by thinking the "unthinkable," namely that square roots of negative numbers could be manipulated in a meaningful way to yield significant results. This was a very bold move on his part. As he put it:

> It was a wild thought in the judgment of many; and I too was for a long time
> of the same opinion. The whole matter seemed to rest on sophistry rather than
> on truth. Yet I sought so long until I actually proved this to be the case [11].

Bombelli developed a "calculus" for complex numbers, stating such rules as $(+\sqrt{-1})(+\sqrt{-1}) = -1$ and $(+\sqrt{-1})(-\sqrt{-1}) = 1$, and defined addition and multiplication of specific complex numbers. This was the birth of complex numbers. But birth did not entail legitimacy. For the next two centuries complex numbers were shrouded in mystery, little understood, and often ignored. Only following their geometric representation in 1831 by Gauss as points in the plane were they accepted as bona fide elements of the number system. (The earlier works of Argand and Wessel on this topic were not well known among mathematicians.) See [1], [7], [13].

Note that the equation $x^3 = 15x + 4$ is an example of an "irreducible cubic," namely one with rational coefficients, irreducible over the rationals, all of whose roots are real. It was shown in the nineteenth century that *any* solution by radicals of such a cubic (not just Cardano's) must involve complex numbers. Thus complex numbers are unavoidable when it comes to finding solutions by radicals of the irreducible cubic. It is for this reason that they arose in connection with the solution of cubic rather than quadratic equations, as is often wrongly assumed. (The nonexistence of a solution of the quadratic $x^2 + 1 = 0$ was readily accepted for centuries.)

1.6 Algebraic notation: Viète and Descartes

Mathematical notation is now taken for granted. In fact, mathematics without a well-developed symbolic notation would be inconceivable. It should be noted, however, that the subject evolved for about three millennia with hardly any symbols. The introduction and perfection of symbolic notation in algebra occurred for the most part in the sixteenth and early seventeenth centuries, and is due mainly to Viète and Descartes.

The decisive step was taken by Viète in his *Introduction to the Analytic Art* (1591). He wanted to breathe new life into the *method of analysis* of the Greeks, a method of discovery used to solve problems, to be contrasted with their method of synthesis, used to prove theorems. The former method he identified with algebra. He saw it as "the science of correct discovery in mathematics," and had the grand vision that it would "leave no problem unsolved."

Viète's basic idea was to introduce arbitrary parameters into an equation and to distinguish these from the equation's variables. He used consonants ($B, C, D, \ldots$) to denote parameters and vowels ($A, E, I, \ldots$) to denote variables. Thus a quadratic equation was written as $BA^2 + CA + D = 0$ (although this was not exactly Viète's notation; see below). To us this appears to be a simple and natural idea, but it was a fundamental departure in algebra: for the first time in over three millennia one could speak of a general quadratic equation, that is, an equation with (arbitrary) literal coefficients rather than one with (specific) numerical coefficients.

François Viète (1540–1603)

This was a seminal contribution, for it transformed algebra from a study of the specific to the general, of equations with numerical coefficients to general equations. Viète himself embarked on a systematic investigation of polynomial equations with literal coefficients. For example, he formulated the relationship between the roots and coefficients of polynomial equations of degree five or less. Algebra had now become considerably more abstract, and was well on its way to becoming a symbolic science.

The ramifications of Viète's work were much broader than its use in algebra. He created a symbolic science that would apply widely, assisting in both the discovery and the demonstration of results. (Compare, for example, Cardano's three-page rhetorical derivation of the formula for the solution of the cubic with the corresponding modern half-page symbolic proof; or, try to discover the relationship between the roots and coefficients of a polynomial equation without the use of symbols.) Viète's ideas proved indispensable in the crucial developments of the seventeenth century—in analytic geometry, calculus, and mathematized science.

His work was not, however, the last word in the formulation of a fully symbolic algebra. The following were some of its drawbacks:

(i) His notation was "syncopated" (i.e., only partly symbolic). For example, an equation such as $x^3 + 3B^2x = 2C^3$ would be expressed by Viète as A cubus + B plano 3 in A aequari C solido 2 (A replaces x here).

(ii) Viète required "homogeneity" in algebraic expressions: all terms had to be of the same degree. That is why the above quadratic is written in what to us is an unusual way, all terms being of the third degree. The requirement of homogeneity goes back to Greek antiquity, where geometry reigned supreme. To the Greek way of thinking, the product ab (say) denoted the *area* of a rectangle with sides a and b; similarly, abc denoted the *volume* of a cube. An expression such as $ab + c$ had no meaning since one could not add length to area. These ideas were an integral part of mathematical practice for close to two millennia.

(iii) Another aspect of the Greek legacy was the geometric justification of algebraic results, as was the case in the works of al-Khwarizmi and Cardano. Viète was no exception in this respect.

(iv) Viète restricted the roots of equations to *positive real* numbers. This is understandable given his geometric bent, for there was at that time no geometric representation for negative or complex numbers.

Most of these shortcomings were overcome by Descartes in his important book *Geometry* (1637), in which he expounded the basic elements of analytic geometry. Descartes' notation was fully symbolic—essentially modern notation (it would be more appropriate to say that modern notation is like Descartes'). For example, he used $x, y, z, \ldots$ for variables and $a, b, c,$ for parameters. Most importantly, he introduced an "algebra of segments." That is, for any two line segments with lengths a and b, he constructed *line segments* with lengths $a + b$, $a - b$, $a \times b$, and a/b. Thus homogeneity of algebraic expressions was no longer needed. For example, $ab + c$ was now a legitimate expression, namely a line segment. This idea represented a most important achievement: it obviated the need for geometry in algebra. For two millennia, geometry had to a large extent been the language of mathematics; now algebra began to play this role. See [1], [7], [10], [12], [17].

1.7 The theory of equations and the Fundamental Theorem of Algebra

Viète's and Descartes' work, in the late sixteenth and early seventeenth centuries, respectively, shifted the focus of attention from the solvability of numerical equations to theoretical studies of equations with literal coefficients. A theory of polynomial equations began to emerge. Among its main concerns were the determination of the existence, nature, and number of roots of such equations. Specifically:

(i) Does every polynomial equation have a root, and, if so, what kind of root is it? This was the most important and difficult of all questions on the subject. It turned out that the first part of the question was much easier to answer than the second. The *Fundamental Theorem of Algebra* (FTA) answered both: every polynomial equation, with real or complex coefficients, has a complex root.

(ii) How many roots does a polynomial equation have? In his *Geometry*, Descartes proved the *Factor Theorem*, namely that if α is a root of the polynomial $p(x)$, then $x - \alpha$ is a factor, that is, $p(x) = (x - \alpha)q(x)$, where $q(x)$ is a polynomial of degree one less than that of $p(x)$. Repeating the process (formally, using induction), it follows that a polynomial of degree n has exactly n roots, given that it has one root, which is guaranteed by the FTA. The n roots need not be distinct. This result means, then, that if $p(x)$ has degree n, there exist n numbers $\alpha_1, \alpha_2, \ldots, \alpha_n$ such that $p(x) = (x - \alpha_1)(x - \alpha_2) \ldots (x - \alpha_n)$. The FTA guarantees that the α_i are complex numbers. (Note that we speak interchangeably of the root α of a *polynomial* $p(x)$ and the root α of a *polynomial equation* $p(x) = 0$; both mean $p(\alpha) = 0$.)

(iii) Can we determine when the roots are rational, real, complex, positive? Every polynomial of *odd* degree with real coefficients has a real root. This was accepted on intuitive grounds in the seventeenth and eighteenth centuries and was formally established in the nineteenth as an easy consequence of the Intermediate Value Theorem in calculus, which says (in the version needed here) that a continuous function $f(x)$ which is positive for some values of x and negative for others, must be zero for some x_0.

Newton showed that the complex roots of a polynomial (if any) appear in conjugate pairs: if $a + bi$ is a root of $p(x)$, so is $a - bi$. Descartes gave an algorithm for finding all *rational* roots (if any) of a polynomial $p(x)$ with *integer* coefficients, as follows. Let $p(x) = a_0 + a_1 x + \cdots + a_n x^n$. If a/b is a rational root of $p(x)$, with a and b relatively prime, then a must be a divisor of a_0 and b a divisor of a_n; since a_0 and a_n have finitely many divisors, this result determines in a finite number of steps all rational roots of $p(x)$ (note that not every a/b for which a divides a_0 and b divides a_n is a rational root of $p(x)$). He also stated (without proof) what came to be known as *Descartes' Rule of Signs*: the number of *positive* roots of a polynomial $p(x)$ does not exceed the number of changes of sign of its coefficients (from "+" to "−" or from "−" to "+"), and the number of *negative* roots is at most the number of times two "+" signs or two "−" signs are found in succession.

(iv) What is the relationship between the roots and coefficients of a polynomial? It had been known for a long time that if α_1 and α_2 are the roots of a quadratic $p(x) = ax^2 + bx + c$, then $\alpha_1 + \alpha_2 = -b/a$ and $\alpha_1 \alpha_2 = c/a$. Viète extended this result to polynomials of degree up to five by giving formulas expressing certain sums and products of the roots of a polynomial in terms of its coefficients. Newton established a general result of this type for polynomials of arbitrary degree, thereby introducing the important notion of *symmetric functions* of the roots of a polynomial.

(v) How do we find the roots of a polynomial? The most desirable way is to determine an exact formula for the roots, preferably a solution by radicals (see the definition on p. 6). We have seen that such formulas were available for polynomials of degree up to four, and attempts were made to extend the results to polynomials of higher degrees (see Chapter 2). In the absence of exact formulas for the roots, various methods were developed for finding *approximate* roots to any desired degree of accuracy. Among the most prominent were Newton's and Horner's methods of the late seventeenth and early nineteenth centuries, respectively. The former involved the use of calculus.

There are several equivalent versions of the Fundamental Theorem of Algebra, including the following:

(i) Every polynomial with complex coefficients has a complex root.
(ii) Every polynomial with real coefficients has a complex root.
(iii) Every polynomial with real coefficients can be written as a product of linear polynomials with complex coefficients.
(iv) Every polynomial with real coefficients can be written as a product of linear and quadratic polynomials with real coefficients.

Statements, but not proofs, of the FTA were given in the early seventeenth century by Girard and by Descartes, although they were hardly as precise as any of the above. For example, Descartes formulated the theorem as follows: "Every equation can have as many distinct roots as the number of dimensions of the unknown quantity in the equation." His "can have" is understandable given that he felt uneasy about the use of complex numbers.

The FTA was important in the calculus of the late seventeenth century, for it enabled mathematicians to find the integrals of rational functions by decomposing their denominators into linear and quadratic factors. But what credence was one to lend to the theorem? Although most mathematicians considered the result to be true, Gottfried Leibniz, for one, did not. For example, in a paper in 1702 he claimed that $x^4 + a^4$ could not be decomposed into linear and quadratic factors.

The first proof of the FTA was given by d'Alembert in 1746, soon to be followed by a proof by Euler. D'Alembert's proof used ideas from analysis (recall that the result was a theorem in algebra), Euler's was largely algebraic. Both were incomplete and lacked rigor, assuming, in particular, that every polynomial of degree n had n roots that could be manipulated according to the laws of the real numbers. What was purportedly proved was that the roots were complex numbers.

Gauss, in his doctoral dissertation completed in 1797 (when he was only twenty years old) and published in 1799, gave a proof of the FTA that was rigorous by the standards of the time. From a modern perspective, Gauss' proof, based on ideas in geometry and analysis, also has gaps. Gauss gave three more proofs (his second and third were essentially algebraic), the last in 1849.

Many proofs of the FTA have since been given, several as recently as the 2000s. Some of them are algebraic, others analytic, and yet others topological. This stands to reason, for a polynomial with complex coefficients is at the same time an algebraic, analytic, and topological object. It is somewhat paradoxical that there is no purely algebraic proof of the FTA: the analytic result that "a polynomial of odd degree over the reals has a real root" has proved to be unavoidable in all algebraic proofs.

In the early nineteenth century the FTA was a relatively new type of result, an *existence theorem*: that is, a mathematical object—a root of a polynomial—was shown to exist, but only in theory. No construction was given for the root. Nonconstructive existence results were very controversial in the nineteenth and early twentieth centuries. Some mathematicians reject them to this day. See [1], [3], [4], [5], [10], [15], [17] for various aspects of this section.

1.8 Symbolical algebra

The study of the solution of polynomial equations inevitably leads to the study of the nature and properties of various number systems, for of course the solutions are themselves numbers. Thus (as we noted) the study of number systems constitutes an important aspect of classical algebra.

The negative and complex numbers, although used frequently in the eighteenth century (the FTA made them inescapable), were often viewed with misgivings and were little understood. For example, Newton described negative numbers as quantities "less than nothing," and Leibniz said that a complex number is "an amphibian between being and nonbeing." Here is Euler on the subject: "We call those *positive quantities*, before which the sign + is found; and those are called *negative quantities*, which are affected by the sign −."

Although rules for the *manipulation* of negative numbers, such as $(-1)(-1) = 1$, had been known since antiquity, no *justification* had in the past been given. (Euler argued that $(-a)(-b)$ must equal ab, for it cannot be $-ab$ since that had been "shown" to be $(-a)b$.) During the late eighteenth and early nineteenth centuries, mathematicians began to ask *why* such rules should hold. Members of the Analytical Society at Cambridge University made important advances on this question. Mathematics at Cambridge was part of liberal arts studies and was viewed as a paradigm of absolute truths employed for the logical training of young minds. It was therefore important, these mathematicians felt, to base algebra, and in particular the laws of operation with negative numbers, on firm foundations.

The most comprehensive work on this topic was Peacock's *Treatise of Algebra* of 1830. His main idea was to distinguish between "arithmetical algebra" and "symbolical algebra." The former referred to operations on symbols that stood only for *positive* numbers and thus, in Peacock's view, needed no justification. For example, $a - (b - c) = a - b + c$ is a law of arithmetical algebra when $b > c$ and $a > b - c$. It becomes a law of symbolical algebra if no restrictions are placed on a, b, and c. In fact, no interpretation of the symbols is called for. Thus *symbolical algebra* was the newly founded subject of operations with symbols that need not refer to specific objects, but that obey the laws of arithmetical algebra. This enabled Peacock to formally establish various laws of algebra. For example, $(-a)(-b)$ was shown to equal ab as follows:

Since $(a-b)(c-d) = ac+bd-ad-bc$ (**) is a law of arithmetical algebra whenever $a > b$ and $c > d$, it becomes a law of symbolical algebra, which holds without restriction on a, b, c, d. Letting $a = 0$ and $c = 0$ in (**) yields $(-b)(-d) = bd$.

Peacock attempted to justify his identification of the laws of symbolical algebra with those of arithmetical algebra by means of the Principle of Permanence of Equivalent Forms, which essentially *decreed* that the laws of symbolical algebra shall be the laws of arithmetical algebra. (What these laws were was not made explicit at the time. They were clarified in the second half of the nineteenth century, when they turned into axioms for rings and fields.) This idea is not very different from the modern approach to the topic in terms of axioms. Its significance was not in the details but in its broad conception, signalling the beginnings of a shift in the essence of algebra from a focus on the meaning of symbols to a stress on their laws of operation. Witness Peacock's description of symbolical algebra:

> In symbolical algebra, the rules determine the meaning of the operations ... we might call them arbitrary assumptions, inasmuch as they are arbitrarily imposed upon a science of symbols and their combinations, which might be adapted to any other assumed system of consistent rules [13].

This was a very sophisticated idea, well ahead of its time. However, Peacock paid only lip service to the arbitrary nature of the laws. In practice, as we have seen, they remained the laws of arithmetic. In the next several decades English mathematicians put into practice what Peacock had preached by introducing algebras with properties which differed in various ways from those of arithmetic. In the words of Bourbaki:

> The algebraists of the English school bring out first, between 1830 and 1850, the abstract notion of law of composition, and enlarge immediately the field of Algebra by applying this notion to a host of new mathematical objects: the algebra of Logic with Boole, vectors, quaternions and general hypercomplex systems with Hamilton, matrices and non-associative laws with Cayley [3].

Thus, whatever its limitations, symbolical algebra provided a positive climate for subsequent developments in algebra. Symbols, and laws of operation on them, began to take on a life of their own, becoming objects of study in their own right rather than a language to represent relationships among numbers. We will see this development in subsequent chapters.

References

1. I. G. Bashmakova and G. S. Smirnova, *The Beginnings and Evolution of Algebra*, The Mathematical Association of America, 2000. (Translated from the Russian by A. Shenitzer.)
2. I. G. Bashmakova and G. S. Smirnova, Geometry: The first universal language of mathematics, in: E. Grosholz and H. Breger (eds), *The Growth of Mathematical Knowledge*, Kluwer, 2000, pp. 331–340.
3. N. Bourbaki, *Elements of the History of Mathematics*, Springer-Verlag, 1991.
4. D. E. Dobbs and R. Hanks, *A Modern Course on the Theory of Equations*, Polygonal Publishing House, 1980.
5. B. Fine and G. Rosenberg, *The Fundamental Theorem of Algebra*, Springer-Verlag, 1987.
6. J. Hoyrup, *Lengths, Widths, Surfaces: A Portrait of Babylonian Algebra and its Kin*, Springer-Verlag, 2002.
7. V. Katz, *A History of Mathematics*, 2nd ed., Addison-Wesley, 1998.
8. V. Katz, Algebra and its teaching: An historical survey, *Journal of Mathematical Behavior* 1997, **16**: 25–38.
9. I. Kleiner, Thinking the unthinkable: The story of complex numbers (with a moral), *Mathematics Teacher* 1988, **81**: 583–592.
10. M. Kline, *Mathematical Thought from Ancient to Modern Times*, Oxford University Press, 1972.
11. P. G. Nahin, *An Imaginary Tale: The Story of $\sqrt{-1}$*, Princeton University Press, 1998.

12. K. H. Parshall, The art of algebra from al-Khwarizmi to Viète: A study in the natural selection of ideas, *History of Science* 1988, **26**: 129–164.

13. H. M. Pycior, George Peacock and the British origins of symbolical algebra, *Historia Mathematica* 1981, **8**: 23–45.

14. E. Robson, Influence, ignorance, or indifference? Rethinking the relationship between Babylonian and Greek mathematics, *Bulletin of the British Society for the History of Mathematics* Spring 2005, **4**: 1–17.

15. H. W. Turnbull, *Theory of Equations*, Oliver and Boyd, 1957.

16. S. Unguru, On the need to rewrite the history of Greek mathematics, *Archive for the History of Exact Sciences* 1975–76, **15**: 67–114.

17. B. L. van der Waerden, *A History of Algebra, from al-Khwarizmi to Emmy Noether*, Springer-Verlag, 1985.

18. B. L. van der Waerden, *Geometry and Algebra in Ancient Civilizations*, Springer-Verlag, 1983.

19. B. L. van der Waerden, Defence of a "shocking" point of view, *Archive for the History of Exact Sciences* 1975–76, **15**: 199–210.

2

History of Group Theory

This chapter will outline the origins of the main concepts, results, and theories discussed in a first course on group theory. These include, for example, the concepts of (abstract) group, normal subgroup, quotient group, simple group, free group, isomorphism, homomorphism, automorphism, composition series, direct product; the theorems of Lagrange, Cauchy, Cayley, Jordan–Hölder; the theories of permutation groups and of abelian groups.

Before dealing with these issues, we wish to mention the context within mathematics as a whole, and within algebra in particular, in which group theory developed. Our "story" concerning the evolution of group theory begins in 1770 and extends to the twentieth century, but the major developments occurred in the nineteenth century. Some of the general mathematical features of that century which had a bearing on the evolution of group theory are: (a) an increased concern for rigor; (b) the emergence of abstraction; (c) the rebirth of the axiomatic method; (d) the view of mathematics as a human activity, possible without reference to, or motivation from, physical situations.

Up to about the end of the eighteenth century algebra consisted, in large part, of the study of solutions of polynomial equations. In the twentieth century algebra became the study of abstract, axiomatic systems. The transition from the so-called classical algebra of polynomial equations to the so-called modern algebra of axiomatic systems occurred in the nineteenth century. In addition to group theory, there emerged the structures of commutative rings, fields, noncommutative rings, and vector spaces. These developed alongside, and sometimes in conjunction with, group theory. Thus Galois theory involved both groups and fields; algebraic number theory contained elements of group theory in addition to commutative ring theory and field theory; group representation theory was a mix of group theory, noncommutative algebra, and linear algebra.

2.1 Sources of group theory

There are four major sources in the evolution of group theory. They are (with the names of the originators and dates of origin):

(a) Classical algebra (Lagrange, 1770)

(b) Number theory (Gauss, 1801)

(c) Geometry (Klein, 1874)

(d) Analysis (Lie, 1874; Poincaré and Klein, 1876)

We deal with each in turn.

2.1.1 Classical Algebra

The major problems in algebra at the time (1770) that Lagrange wrote his fundamental memoir "Reflections on the solution of algebraic equations" concerned polynomial equations. There were "theoretical" questions dealing with the existence and nature of the roots—for example, does every equation have a root? how many roots are there? are they real, complex, positive, negative?—and "practical" questions dealing with methods for finding the roots. In the latter instance there were exact methods and approximate methods. In what follows we mention exact methods.

The Babylonians knew how to solve quadratic equations, essentially by the method of completing the square, around 1600 BC (see Chapter 1). Algebraic methods for solving the cubic and the quartic were given around 1540 (Chapter 1). One of the major problems for the next two centuries was the algebraic solution of the quintic. This is the task Lagrange set for himself in his paper of 1770.

Joseph Louis Lagrange (1736–1813)

In this paper Lagrange first analyzed the various known methods, devised by Viète, Descartes, Euler, and Bezout, for solving cubic and quartic equations. He showed that the common feature of these methods is the reduction of such equations to auxiliary equations—the so-called *resolvent equations*. The latter are one degree lower than the original equations.

Lagrange next attempted a similar analysis of polynomial equations of arbitrary degree n. With each such equation he associated a resolvent equation, as follows: let $f(x)$ be the original equation, with roots $x_1, x_2, x_3, \ldots, x_n$. Pick a rational function $\mathbf{R}(x_1, x_2, x_3, \ldots, x_n)$ of the roots and coefficients of $f(x)$. (Lagrange described methods for doing this.) Consider the different values which $\mathbf{R}(x_1, x_2, x_3, \ldots, x_n)$ assumes under all the $n!$ permutations of the roots $x_1, x_2, x_3, \ldots, x_n$ of $f(x)$. If these are denoted by $y_1, y_2, y_3, \ldots, y_k$, then the resolvent equation is given by $g(x) = (x - y_1)(x - y_2) \cdots (x - y_k)$.

It is important to note that the coefficients of $g(x)$ are symmetric functions in $x_1, x_2, x_3, \ldots, x_n$, hence they are polynomials in the elementary symmetric functions of $x_1, x_2, x_3, \ldots, x_n$; that is, they are polynomials in the coefficients of the original equation $f(x)$. Lagrange showed that k divides $n!$—the source of what we call *Lagrange's theorem* in group theory.

For example, if $f(x)$ is a quartic with roots x_1, x_2, x_3, x_4, then $\mathbf{R}(x_1, x_2, x_3, x_4)$ may be taken to be $x_1 x_2 + x_3 x_4$, and this function assumes three distinct values under the twenty-four permutations of x_1, x_2, x_3, x_4. Thus the resolvent equation of a quartic is a cubic. However, in carrying over this analysis to the quintic Lagrange found that the resolvent equation is of degree six.

Although Lagrange did not succeed in resolving the problem of the algebraic solvability of the quintic, his work was a milestone. It was the first time that an association was made between the solutions of a polynomial equation and the permutations of its roots. In fact, the study of the permutations of the roots of an equation was a cornerstone of Lagrange's general theory of algebraic equations. This, he speculated, formed "the true principles of the solution of equations." He was, of course, vindicated in this by Galois. Although Lagrange spoke of permutations without considering a "calculus" of permutations (e.g., there is no consideration of their composition or closure), it can be said that the germ of the group concept—as a group of permutations—is present in his work. For details see [12], [16], [19], [25], [33].

2.1.2 Number Theory

In the *Disquisitiones Arithmeticae (Arithmetical Investigations)* of 1801 Gauss summarized and unified much of the number theory that preceded him. The work also suggested new directions which kept mathematicians occupied for the entire century. As for its impact on group theory, the *Disquisitiones* may be said to have initiated the theory of finite abelian groups. In fact, Gauss established many of the significant properties of these groups without using any of the terminology of group theory.

The groups appeared in four different guises: the additive group of integers modulo m, the multiplicative group of integers relatively prime to m, modulo m, the group of equivalence classes of binary quadratic forms, and the group of n-th roots of unity. And although these examples turned up in number-theoretic contexts, it is as abelian groups that Gauss treated them, using what are clear prototypes of modern algebraic proofs.

For example, considering the nonzero integers modulo p (p a prime), he showed that they are all powers of a single element; that is, that the group Z_p^* of such integers is cyclic. Moreover, he determined the number of generators of this group, showing that it is equal to $\varphi(p-1)$, where φ is Euler's φ-function.

Given any element of Z_p^*, he defined the order of the element (without using the terminology) and showed that the order of an element is a divisor of $p-1$. He then used this result to prove Fermat's "little theorem," namely that $a^{p-1} \equiv 1 \pmod{p}$ if p does not divide a, thus employing group-theoretic ideas to prove number-theoretic results. Next he showed that if t is a positive integer which divides $p-1$, then there exists an element in Z_p^* whose order is t—essentially the converse of Lagrange's theorem for cyclic groups.

Concerning the n-th roots of 1, which he considered in connection with the cyclotomic equation, he showed that they too form a cyclic group. In relation to this group he raised and answered many of the same questions he raised and answered in the case of Z_p^*.

The problem of representing integers by binary quadratic forms goes back to Fermat in the early seventeenth century. (Recall his theorem that every prime of the form $4n + 1$ can be represented as a sum of two squares $x^2 + y^2$.) Gauss devoted a large part of the *Disquisitiones* to an exhaustive study of binary quadratic forms and the representation of integers by such forms.

A binary quadratic form is an expression of the form $ax^2 + bxy + cy^2$, with a, b, c integers. Gauss defined a composition on such forms, and remarked that if K_1 and K_2 are two such forms, one may denote their composition by $K_1 + K_2$. He then showed that this composition is associative and commutative, that there exists an identity, and that each form has an inverse, thus verifying all the properties of an abelian group.

Despite these remarkable insights, one should not infer that Gauss had the concept of an abstract group, or even of a finite abelian group. Although the arguments in the *Disquisitiones* are quite general, each of the various types of "groups" he considered was dealt with separately—there was no unifying group-theoretic method which he applied to all cases.

For further details see [5], [9], [25], [30], [33].

2.1.3 Geometry

We are referring here to Klein's famous and influential (but see [18]) lecture entitled "A Comparative Review of Recent Researches in Geometry," which he delivered in 1872 on the occasion of his admission to the faculty of the University of Erlangen. The aim of this so-called Erlangen Program was the classification of geometry as the study of invariants under various groups of transformations. Here there appear groups such as the projective group, the group of rigid motions, the group of similarities, the hyperbolic group, the elliptic groups, as well as the geometries associated with them. (The affine group was not mentioned by Klein.) Now for some background leading to Klein's Erlangen Program.

The nineteenth century witnessed an explosive growth in geometry, both in scope and in depth. New geometries emerged: projective geometry, noneuclidean geometries, differential geometry, algebraic geometry, n-dimensional geometry, and

Grassmann's geometry of extension. Various geometric methods competed for supremacy: the synthetic versus the analytic, the metric versus the projective.

At mid-century a major problem had arisen, namely the classification of the relations and inner connections among the different geometries and geometric methods. This gave rise to the study of "geometric relations," focusing on the study of properties of figures invariant under transformations. Soon the focus shifted to a study of the transformations themselves. Thus the study of the geometric relations of figures became the study of the associated transformations.

Various types of transformations (e.g., collineations, circular transformations, inversive transformations, affinities) became the objects of specialized studies. Subsequently, the logical connections among transformations were investigated, and this led to the problem of classifying transformations, and eventually to Klein's group-theoretic synthesis of geometry.

Klein's use of groups in geometry was the final stage in bringing order to geometry. An intermediate stage was the founding of the first major theory of classification in geometry, beginning in the 1850s, the Cayley–Sylvester Invariant Theory. Here the objective was to study invariants of "forms" under transformations of their variables (see Chapter 8.1). This theory of classification, the precursor of Klein's Erlangen Program, can be said to be implicitly group-theoretic. Klein's use of groups in geometry was, of course, explicit. (For a thorough analysis of implicit group-theoretic thinking in geometry leading to Klein's Erlangen Program see [33].)

In the next section we will note the significance of Klein's Erlangen Program (and his other works) for the evolution of group theory. Since the Program originated a hundred years after Lagrange's work and eighty years after Gauss' work, its importance for group theory can best be appreciated after a discussion of the evolution of group theory beginning with the works of Lagrange and Gauss and ending with the period around 1870.

2.1.4 Analysis

In 1874 Lie introduced his general theory of continuous transformation groups—essentially what we call Lie groups today. Such a group is represented by the transformations $x_i^1 = f_i(x_1, x_2, \ldots, x_n, a_1, a_2, \ldots, a_n), i = 1, 2, \ldots, n$, where the f_i are analytic functions in the x_i and a_i (the a_i are parameters, with both x_i and a_i real or complex). For example, the transformations given by $x^1 = (ax+b)/(cx+d)$, where a, b, c, d are real numbers and $ad - bc \neq 0$, define a continuous transformation group.

Lie thought of himself as the successor of Abel and Galois, doing for differential equations what they had done for algebraic equations. His work was inspired by the observation that almost all the differential equations which had been integrated by the older methods remain invariant under continuous groups that can be easily constructed. He was then led to consider, in general, differential equations that remain invariant under a given continuous group and to investigate the possible simplifications in these equations which result from the known properties of the given group (cf. Galois theory). Although Lie did not succeed in the actual formulation of a

"Galois theory of differential equations," his work was fundamental in the subsequent formulation of such a theory by Picard (1883-1887) and Vessiot (1892).

Poincaré and Klein began their work on "automorphic functions" and the groups associated with them around 1876. Automorphic functions (which are generalizations of the circular, hyperbolic, elliptic, and other functions of elementary analysis) are functions of a complex variable z, analytic in some domain D, which are invariant under the group of transformations $x^1 = (ax+b)/(cx+d)$ (a, b, c, d real or complex and $ad - bc \neq 0$), or under some subgroup of this group. Moreover, the group in question must be "discontinuous," that is, any compact domain contains only finitely many transforms of any point.

Examples of such groups are the modular group (in which a, b, c, d are integers and $ad - bc = 1$), which is associated with the elliptic modular functions, and Fuchsian groups (in which a, b, c, d are real and $ad - bc = 1$) associated with the Fuchsian automorphic functions. As in the case of Klein's Erlangen Program, we will explore the consequences of these works for group theory in the next section.

2.2 Development of "specialized" theories of groups

In the previous section we outlined four major sources in the evolution of group theory. The first source—classical algebra—led to the theory of permutation groups; the second source—number theory—led to the theory of abelian groups; the third and fourth sources—geometry and analysis—led to the theory of transformation groups. We will now outline some developments within these specialized theories.

2.2.1 Permutation Groups

As noted earlier, Lagrange's work of 1770 initiated the study of permutations in connection with the study of the solution of equations. It was probably the first clear instance of implicit group-theoretic thinking in mathematics. It led directly to the works of Ruffini, Abel, and Galois during the first third of the nineteenth century, and to the concept of a permutation group.

Ruffini and Abel proved the unsolvability of the quintic by building on the ideas of Lagrange concerning resolvents. Lagrange showed that a necessary condition for the solvability of the general polynomial equation of degree n is the existence of a resolvent of degree less than n. Ruffini and Abel showed that such resolvents do not exist for $n > 4$. In the process they developed elements of permutation theory. It was Galois, however, who made the fundamental conceptual advances, and who is considered by many as the founder of (permutation) group theory.

He was familiar with the works of Lagrange, Abel, and Gauss on the solution of polynomial equations. But his aim went well beyond finding a method for solvability of equations. He was concerned with gaining insight into general principles, dissatisfied as he was with the methods of his predecessors: "From the beginning of this century," he wrote, "computational procedures have become so complicated that any progress by those means has become impossible."

Galois recognized the separation between "Galois theory"—the correspondence between fields and groups—and its application to the solution of equations, for he wrote that he was presenting "the general principles and just one application" of the theory. "Many of the early commentators on Galois theory failed to recognize this distinction, and this led to an emphasis on applications at the expense of the theory" [19].

Galois was the first to use the term "group" in a technical sense—to him it signified a collection of permutations closed under multiplication: "If one has in the same group the substitutions S and T, one is certain to have the substitution ST." He recognized that the most important properties of an algebraic equation were reflected in certain properties of a group uniquely associated with the equation—"the group of the equation." To describe these properties he invented the fundamental notion of normal subgroup and used it to great effect.

While the issue of resolvent equations preoccupied Lagrange, Ruffini, and Abel, Galois' basic idea was to bypass them, for the construction of a resolvent required great skill and was not based on a clear methodology. Galois noted instead that the existence of a resolvent was equivalent to the existence of a normal subgroup of prime index in the group of the equation. This insight shifted consideration from the resolvent equation to the group of the equation and its subgroups.

Galois defined the group of an equation as follows:

Let an equation be given, whose m roots are $a, b, c, \ldots$. There will always be a group of permutations of the letters $a, b, c, \ldots$ which has the following property: (1) that every function of the roots, invariant under the substitutions of that group, is rationally known [i.e., is a rational function of the coefficients and any adjoined quantities]; (2) conversely, that every function of the roots, which can be expressed rationally, is invariant under these substitutions [19].

The definition says essentially that the group of an equation consists of those permutations of the roots of the equation which leave invariant all relations among the roots over the field of coefficients of the equation—basically the definition we would give today. Of course the definition does not guarantee the existence of such a group, and so Galois proceeded to demonstrate it. He next investigated how the group changes when new elements are adjoined to the "ground field." His treatment was close to the standard treatment of this matter in a modern algebra text.

Galois' work was slow in being understood and assimilated. In fact, while it was done around 1830, it was published posthumously by Liouville, in 1846. Beyond his technical accomplishments,

Galois forced later developments substantially and in two ways. On the one hand, since he discovered theorems for which he could not give proofs based on clearly defined concepts and calculations, it was inevitable that his successors would find it necessary to fill the gaps. On the other hand, it would not be enough merely to prove the correctness of these theorems; their substance, their group-theoretic core, must be extracted [33].

For details see [12], [19], [23], [25], [29], [31], [33]. See also Chapter 8.3.

The other major contributor to permutation theory in the first half of the nineteenth century was Cauchy. In several major papers in 1815 and 1844 he inaugurated the theory of permutation groups as an autonomous subject. Before Cauchy, permutations were not an object of independent study but rather a useful device for the investigation of solutions of polynomial equations. Although Cauchy was well aware of the work of Lagrange and Ruffini (Galois' work was not yet published at the time), he "was definitely not inspired directly by the contemporary group-theoretic formulation of the problem of solvability of algebraic equations" [33].

Augustin-Louis Cauchy (1789–1857)

In these works Cauchy gave the first systematic development of the subject of permutation groups. In the 1815 papers he used no special name for sets of permutations closed under multiplication. However, he recognized their importance and gave a name to the number of elements in such a closed set, calling it "diviseur indicatif." In the 1844 paper he defined the concept of a group of permutations generated by certain elements:

Given one or more substitutions involving some or all of the elements $x, y, z, \ldots$, I call the products of these substitutions, by themselves or by any other, in any order, *derived* substitutions. The given substitutions, together with the derived ones, form what I call a *system of conjugate substitutions* [22].

In these papers, which were very influential, Cauchy made several lasting additions to the terminology, notation, and results of permutation theory. For example, he introduced the permutation notation in use today, as well as the cyclic notation for permutations; defined the product of permutations, the degree of a permutation, cyclic permutation, and transposition; recognized the identity permutation as a permutation; discussed what we would call today the direct product of two groups; and dealt with the alternating groups extensively. Here is a sample of some of the results he proves.

(i)   Every even permutation is a product of 3-cycles.
(ii) If p (prime) is a divisor of the order of a group, there exists a subgroup of order p. This is known today as *Cauchy's theorem*, though it was stated without proof by Galois.
(iii) Determines all subgroups of S_3, S_4, S_5, and S_6 (making an error in S_6).
(iv) All permutations which commute with a given one form a group, nowadays called the *centralizer* of an element of the group.

It should be noted that all these results were given and proved in the context of permutation groups. For details see [6], [8], [23], [24], [25], [33].

The crowning achievement of these two lines of development—a symphony on the grand themes of Galois and Cauchy—was Jordan's important and influential *Treatise on Substitutions and Algebraic Equations* of 1870. Although the author stated in the preface that "the aim of the work is to develop Galois' method and to make it a proper field of study, by showing with what facility it can solve all principal problems of the theory of equations," it is in fact group theory per se—not as an offshoot of the theory of solvability of equations—which formed the central object of study.

The striving for a mathematical synthesis based on key ideas is a striking characteristic of Jordan's work as well as that of a number of other mathematicians of the period, for example Klein. The concept of a (permutation) group seemed to Jordan to provide such a key idea. His approach enabled him to give a unified presentation of results due to Galois, Cauchy, and others. His application of the group concept to the theory of equations, algebraic geometry, transcendental functions, and theoretical mechanics was also part of the unifying and synthesizing theme. "In his book Jordan wandered through all of algebraic geometry, number theory, and function theory in search of interesting permutation groups" [20]. In fact, his aim was a survey of all of mathematics by areas in which the theory of permutation groups had been applied or seemed likely to be applicable. "The work represents ... a review of the whole of contemporary mathematics from the standpoint of the occurrence of group-theoretic thinking in permutation-theoretic form" [33].

The *Treatise* embodied the substance of most of Jordan's publications on groups up to that time (he wrote over 30 articles on groups during the period 1860—1880) and directed attention to a large number of difficult problems, introducing many fundamental concepts. For example, he made explicit the notions of *isomorphism* and *homomorphism* for (substitution) groups, introduced the term "solvable group" for the first time in a technical sense, introduced the concept of a *composition series*, and proved part of the *Jordan–Hölder theorem*, namely that the indices in two composition series are the same (the concept of a quotient group was not explicitly recognized at this time); and he undertook a very thorough study of transitivity and primitivity for permutation groups, obtaining results most of which have not since been superseded. He also gave a proof that A_n is simple for $n > 4$.

An important part of the treatise was devoted to a study of the "linear group" and some of its subgroups. In modern terms these constitute the so-called classical groups, namely the general linear group, the unimodular group, the orthogonal group, and the symplectic group. Jordan considered these groups only over finite fields, and proved their simplicity in certain cases. It should be noted, however, that he took these groups to be permutation groups rather than groups of matrices or linear transformations.

Jordan's *Treatise* is a landmark in the evolution of group theory. His permutation-theoretic point of view, however, was soon to be overtaken by the conception of a group as a group of transformations (see 2.2.3 below). "The *Traité* [*Treatise*] marks a break in the evolution and application of the permutation-theoretic group concept. It was an expression of Jordan's deep desire to bring about a conceptual synthesis of the mathematics of his time. That he tried to achieve such a synthesis by relying on the concept of a permutation group, which the very next phase of mathematical development would show to have been unduly restricted, makes for both the glory and the limitations of his book ..." [33]. For details see [9], [13], [19], [20], [22], [24], [29], [33].

2.2.2 Abelian Groups

As noted earlier, the main source for abelian group theory was number theory, beginning with Gauss' *Disquisitiones Arithmeticae*. (Note also implicit abelian group theory in Euler's number-theoretic work [33].) In contrast to permutation theory, group-theoretic modes of thought in number theory remained implicit until about the last third of the nineteenth century. Until that time no explicit use of the term "group" was made, and there was no link to the contemporary, flourishing theory of permutation groups. We now give a sample of some implicit group-theoretic work in number theory, especially in algebraic number theory.

Algebraic number theory arose in connection with Fermat's Last Theorem, the insolvability in nonzero integers of $x^n + y^n = z^n$ for $n > 2$, Gauss' theory of binary quadratic forms, and higher reciprocity laws (see Chapter 3.2). Algebraic number fields and their arithmetical properties were the main objects of study. In 1846 Dirichlet studied the units in an algebraic number field and established that (in our terminology) the group of these units is a direct product of a finite cyclic group and a free abelian group of finite rank. At about the same time Kummer introduced his "ideal numbers," defined an equivalence relation on them, and derived, for cyclotomic fields, certain special properties of the number of equivalence classes, the so-called class number of a cyclotomic field—in our terminology, the order of the ideal class group of the cyclotomic field. Dirichlet had earlier made similar studies of quadratic fields.

In 1869 Schering, a former student of Gauss, investigated the structure of Gauss' (group of) equivalence classes of binary quadratic forms (see Chapter 3). He found certain fundamental classes from which all classes of forms could be obtained by composition. In group-theoretic terms, Schering found a basis for the abelian group of equivalence classes of binary quadratic forms.

Kronecker generalized Kummer's work on cyclotomic fields to arbitrary algebraic number fields. In an 1870 paper on algebraic number theory, entitled "An exposition of some properties of the class number of ideal complex numbers," he began by taking a very abstract point of view. He considered a finite set of arbitrary "elements," and defined an abstract operation on them which satisfied certain laws—laws which we may nowadays take as axioms for a finite abelian group:

Let $\theta^1, \theta^{11}, \theta^{111}$ be finitely many elements such that with any two of them we can associate a third by means of a definite procedure. Thus, if f denotes the procedure and θ^1, θ^{11} are two (possibly equal) elements, then there exists a θ^{111} equal to $f(\theta^1, \theta^{11})$. Furthermore, $f(\theta^1, \theta^{11}) = f(\theta^{11}, \theta^1)$, $f(\theta^1, f(\theta^{11}, \theta^{111})) = f(f(\theta^1, \theta^{11}), \theta^{111})$, and if θ^{11} is different from θ^{111} then $f(\theta^1, \theta^{11})$ is different from $f(\theta^1, \theta^{111})$. Once this is assumed we can replace the operation $f(\theta^1, \theta^{11})$ by multiplication $\theta^1 \cdot \theta^{11}$ provided that instead of equality we employ equivalence. Thus using the usual equivalence symbol "$\sim$" we define the equivalence $\theta^1 \cdot \theta^{11} \sim \theta^{111}$ by means of the equation $f(\theta^1, \theta^{11}) = \theta^{111}$ [33].

Kronecker aimed at working out the laws of combination of "magnitudes," in the process giving an implicit definition of a finite abelian group. From the above abstract considerations Kronecker deduced the following consequences:

(i) If θ is any "element" of the set under discussion, then $\theta^k = 1$ for some positive integer k. If k is the smallest such, then θ is said to "belong to k." If θ belongs to k and $\theta^m = 1$, then k divides m.

(ii) If an element θ belongs to k, then every divisor of k has an element belonging to it.

(iii) If θ and θ^1 belong to k and k^1 respectively, and k and k^1 are relatively prime, then $\theta\theta^1$ belongs to kk^1.

(iv) There exists a "fundamental system" of elements $\theta_1, \theta_2, \theta_3, \ldots$ such that the expression $\theta_1^{h_1}\theta_2^{h_2}\theta_3^{h_3} \ldots (h_i = 1, 2, 3, \ldots, n_i)$ represents each element of the given set of elements just once. The numbers $n_1, n_2, n_3, \ldots$ to which, respectively, $\theta_1, \theta_2, \theta_3, \ldots$ belong, are such that each is divisible by its successor; the product $n_1 n_2 n_3 \ldots$ is equal to the totality of elements of the set.

The above can, of course, be interpreted as well-known results on finite abelian groups; in particular, (iv) can be taken as the *basis theorem* for such groups. Once Kronecker established this general framework, he applied it to the special cases of equivalence classes of binary quadratic forms and to ideal classes. He noted that when applying (iv) to the former one obtains Schering's result (see p. 26).

Although Kronecker did not relate his implicit definition of a finite abelian group to the by that time well-established concept of a permutation group, of which he was well aware, he clearly recognized the advantages of the abstract point of view which he adopted:

The very simple principles ... are applied not only in the context indicated but also frequently elsewhere—even in the elementary parts of number theory.

This shows, and it is otherwise easy to see, that these principles belong to a more general and more abstract realm of ideas. It is therefore proper to free their development from all inessential restrictions, thus making it unnecessary to repeat the same argument when applying it in different cases. ... Also, when stated with all admissible generality, the presentation gains in simplicity and, since only the truly essential features are thrown into relief, in transparency [33].

The above lines of development were capped in 1879 by an important paper of Frobenius and Stickelberger entitled "On groups of commuting elements." Although they built on Kronecker's work, they used the concept of an abelian group explicitly and, moreover, made the important advance of recognizing that the abstract group concept embraces congruences and Gauss' composition of forms as well as the substitution groups of Galois. They also mentioned, in footnotes, groups of infinite order, namely groups of units of number fields and the group of all roots of unity. One of their main results was a proof of the *basis theorem for finite abelian groups*, including a proof of the uniqueness of decomposition. It is interesting to compare their explicit, "modern" formulation of the theorem to that of Kronecker ((iv) above):

A group that is not irreducible [indecomposable] can be decomposed into purely irreducible factors. As a rule, such a decomposition can be accomplished in many ways. However, regardless of the way in which it is carried out, the number of irreducible factors is always the same and the factors in the two decompositions can be so paired off that the corresponding factors have the same order [33].

They went on to identify the "irreducible factors" as cyclic groups of prime power orders. They then applied their results to groups of integers modulo m, binary quadratic forms, and ideal classes in algebraic number fields.

The paper by Frobenius and Stickelberger is "a remarkable piece of work, building up an independent theory of finite abelian groups on its own foundation in a way close to modern views" [30]. For details on this section see [5], [9], [24], [30], [33].

2.2.3 Transformation Groups

As in number theory, so in geometry and analysis, group-theoretic ideas remained implicit until the last third of the nineteenth century. Moreover, Klein's (and Lie's) explicit use of groups in geometry influenced the evolution of group theory conceptually rather than technically. It signified a genuine shift in the development of group theory from a preoccupation with permutation groups to the study of groups of transformations. (That is not to suggest, of course, that permutation groups were no longer studied.) This transition was also notable in that it pointed to a turn from finite groups to infinite groups.

Klein noted the connection of his work with permutation groups but also realized the departure he was making. He stated that what Galois theory and his own program have in common is the investigation of "groups of changes," but added that "to be sure,

Felix Klein (1849–1925)

the objects the changes apply to are different: there [Galois theory] one deals with a finite number of discrete elements, whereas here one deals with an infinite number of elements of a continuous manifold." To continue the analogy, Klein noted that just as there is a theory of permutation groups, "we insist on a *theory of transformations*, a study of groups generated by transformations of a given type."

Klein shunned the abstract point of view in group theory, and even his technical definition of a (transformation) group is deficient:

> Now let there be given a sequence of transformations $A, B, C, \ldots$. If this sequence has the property that the composite of any two of its transformations yields a transformation that again belongs to the sequence, then the latter will be called a group of transformations [33].

Klein's work, however, broadened considerably the conception of a group and its applicability in other fields of mathematics. He did much to promote the view that group-theoretic ideas are fundamental in mathematics:

> The special subject of group theory extends through all of modern mathematics. As an ordering and classifying principle, it intervenes in the most varied domains.

There was another context in which groups were associated with geometry, namely "motion geometry," that is, the use of motions or transformations of geometric objects as group elements. Already in 1856 Hamilton considered (implicitly) "groups" of the regular solids. Jordan, in 1868, dealt with the classification of all subgroups of the group of motions of Euclidean 3-space. And Klein in his *Lectures on the Icosahedron* of 1884 "solved" the quintic equation by means of the symmetry group of the icosahedron. He thus discovered a deep connection between the groups of rotations of the regular solids, polynomial equations, and complex function theory. In these *Lectures* there also appeared the "Klein 4-group."

In the late 1860s Klein and Lie had jointly undertaken "to investigate geometric or analytic objects that are transformed into themselves by *groups of changes*." (This is Klein's retrospective description, in 1894, of their program.) While Klein concentrated on discrete groups, Lie studied continuous transformation groups. Lie realized that the theory of continuous transformation groups was a very powerful tool in geometry and differential equations and he set himself the task of "determining all groups of . . . [continuous] transformations." He achieved his objective by the early 1880s with the classification of these groups. A classification of discontinuous transformation groups was obtained by Poincaré and Klein a few years earlier.

Beyond the technical accomplishments in the areas of discontinuous and continuous transformation groups—extensive theories developed in both areas and both are still active fields of research, what is important for us in the founding of these theories is the following:

(i) They provided a major extension of the scope of the concept of a group—from permutation groups and abelian groups to transformation groups;
(ii) They introduced important examples of infinite groups—previously the only objects of study were finite groups;
(iii) They greatly extended the range of applications of the group concept to include number theory, the theory of algebraic equations, geometry, the theory of differential equations—both ordinary and partial, and function theory (automorphic functions, complex functions).

All this occurred prior to the emergence of the abstract group concept. In fact, these developments were instrumental in the emergence of the concept of an abstract group, which we describe next. For further details on this section see [5], [7], [9], [17], [18], [20], [24], [29], [33].

2.3 Emergence of abstraction in group theory

The abstract point of view in group theory emerged slowly. It took over one hundred years from the time of Lagrange's implicit group-theoretic work of 1770 for the abstract group concept to evolve. E. T. Bell discerns several stages in this process of evolution towards abstraction and axiomatization:

> The entire development required about a century. Its progress is typical of the evolution of any major mathematical discipline of the recent period; first, the discovery of isolated phenomena; then the recognition of certain features common to all; next the search for further instances, their detailed calculation and classification; then the emergence of general principles making further calculations, unless needed for some definite application, superfluous; and last, the formulation of postulates crystallizing in abstract form the structure of the system investigated [2].

Although somewhat oversimplified, as all such generalizations tend to be, this is nevertheless a useful framework. Indeed, in the case of group theory, first came the "isolated phenomena"—for example, permutations, binary quadratic forms, roots of unity; then the recognition of "common features"—the concept of a finite group, encompassing both permutation groups and finite abelian groups (cf. the paper of Frobenius and Stickelberger cited above); next the search for "other instances"—in our case transformation groups; and finally the formulation of "postulates"—in this case the postulates of a group, encompassing both the finite and infinite cases. We now consider when and how the intermediate and final stages of abstraction occurred.

In 1854 Cayley gave the first abstract definition of a finite group in a paper entitled "On the theory of groups, as depending on the symbolic equation $\theta^n = 1$." (In 1858 Dedekind, in lectures on Galois theory at Göttingen, gave another. See 8.2.) Here is Cayley's definition:

> A set of symbols $1, \alpha, \beta, \ldots$, all of them different, and such that the product of any two of them (no matter in what order), or the product of any one of them into itself, belongs to the set, is said to be a *group*.

Cayley went on to say that:

> These symbols are not in general convertible [commutative] but are associative ... and it follows that if the entire group is multiplied by any one of the symbols, either as further or nearer factor [i.e., on the left or on the right], the effect is simply to reproduce the group [33].

He then presented several examples of groups, such as the quaternions (under addition), invertible matrices (under multiplication), permutations, Gauss' quadratic forms, and groups arising in elliptic function theory. Next he showed that every abstract group is (in our terminology) isomorphic to a permutation group, a result now known as *Cayley's theorem*.

He seemed to have been well aware of the concept of *isomorphic* groups, although he did not define it explicitly. However, he introduced the multiplication table of a (finite) group and asserted that an abstract group is determined by its multiplication table. He then determined all the groups of orders four and six, showing there are two of each by displaying multiplication tables. Moreover, he noted that the cyclic group of order n "is in every respect analogous to the system of the roots of the ordinary equation $x^n - 1 = 0$," and that there exists only one group of a given prime order. See [35] for a discussion of Cayley's definition of an abstract group. See also Chapter 8.1.

Cayley's orientation towards an abstract view of groups—a remarkable accomplishment at this time in the evolution of group theory—was due, at least in part, to his contact with the abstract work of Boole. The concern with the abstract foundations of mathematics was characteristic of the circles around Boole, Cayley, and Sylvester already in the 1840s.

Cayley's achievement was, however, only a personal triumph. His abstract definition of a group attracted no attention at the time, even though Cayley was already well known. The mathematical community was apparently not ready for such abstraction: permutation groups were the only groups under serious investigation, and more generally, the formal approach to mathematics was still in its infancy. "Premature abstraction falls on deaf ears, whether they belong to mathematicians or to students," as Kline put it in his inimitable way [21].

For further details see [22], [23], [24], [25], [29], [33].

It was only a quarter of a century later that the abstract group concept began to take hold. And it was Cayley again who in four short papers on group theory written in 1878 returned to the abstract point of view he adopted in 1854. Here he stated the general problem of finding all groups of a given order and showed that any (finite) group is isomorphic to a group of permutations. But, as he remarked:

> This ... does not in any wise show that the best or easiest mode of treating the general problem is thus to regard it as a problem of substitutions: and it seems clear that the better course is to consider the general problem in itself, and to deduce from it the theory of groups of substitutions.

These papers of Cayley, unlike those of 1854, inspired a number of fundamental group-theoretic works.

Another mathematician who advanced the abstract point of view in group theory (and more generally in algebra) was Weber. Here is his "modern" definition of an abstract (finite) group given in an 1882 paper on quadratic forms [23]:

> A system G of h arbitrary elements $\theta_1, \theta_2, \ldots, \theta_h$ is called a group of degree h if it satisfies the following conditions:
>
> I. By some rule which is designated as composition or multiplication, from any two elements of the same system one derives a new element of the same system. In symbols, $\theta_r \theta_s = \theta_t$.
>
> II. It is always true that $(\theta_r \theta_s)\theta_t = \theta_r(\theta_s \theta_t) = \theta_r \theta_s \theta_t$.
>
> III. From $\theta \theta_r = \theta \theta_s$ or from $\theta_r \theta = \theta_s \theta$ it follows that $\theta_r = \theta_s$.

Weber's and other definitions of abstract groups given at the time applied only to finite groups. They thus encompassed the two theories of permutation groups and (finite) abelian groups, which derived from the two sources of classical algebra—polynomial equations and number theory, respectively. Infinite groups, which arose from the theories of (discontinuous and continuous) transformation groups, were not subsumed under those definitions.

It was W. von Dyck who, in an important and influential paper in 1882 entitled "Group-theoretic studies," consciously included and combined, for the first time, all of the major historical roots of abstract group theory—the algebraic, number-theoretic, geometric, and analytic. In his own words:

> The following investigations aim to continue the study of the properties of a group in its abstract formulation. In particular, this will pose the question of the extent to which these properties have an invariant character present in all the different realizations of the group, and the question of what leads to the exact determination of their essential group-theoretic content [33].

Von Dyck's definition of an abstract group, which included both the finite and infinite cases, was given in terms of generators (he calls them "operations") and defining relations (the definition is somewhat long—see [7]). He stressed that "in this way all ... isomorphic groups are included in a single group," and that "the essence of a group is no longer expressed by a particular form of its operations but rather by their mutual relations." He then went on to construct the *free group* on n generators, and showed (essentially, without using the terminology) that every finitely generated group is a quotient group of a free group of finite rank.

What is important from the point of view of postulates for group theory is that von Dyck was the first to require explicitly the existence of an inverse in his definition of a group: "We require for our considerations that a group which contains the operation T_k must also contain its inverse T_k^{-1}." In a second paper (in 1883) von Dyck applied his abstract development of group theory to permutation groups, finite rotation groups (symmetries of polyhedra), number-theoretic groups, and transformation groups.

Although various postulates for groups appeared in the mathematical literature for the next twenty years, the abstract point of view in group theory was not universally applauded. In particular, Klein, one of the major contributors to the development of group theory, thought that the "abstract formulation is excellent for the working out of proofs but it does not help one find new ideas and methods," adding that "in general, the disadvantage of the [abstract] method is that it fails to encourage thought" [33].

Despite Klein's reservations, the mathematical community was at this time (early 1880s) receptive to the abstract formulations (cf. the response to Cayley's definition of 1854). The major reasons for this receptivity were:

(i) There were now several major "concrete" theories of groups—permutation groups, abelian groups, discontinuous transformation groups (the finite and infinite cases), and continuous transformation groups, and this warranted abstracting their essential features.

(ii) Groups came to play a central role in diverse fields of mathematics, such as different parts of algebra, geometry, number theory, and several areas of analysis, and the abstract view of groups was thought to clarify what was essential for such applications and to offer opportunities for further applications.

(iii) The formal approach, aided by the penetration into mathematics of set theory and mathematical logic, became prevalent in other fields of mathematics, for example, various areas of geometry and analysis.

In the next section we will follow, very briefly, the evolution of that abstract point of view in group theory.

2.4 Consolidation of the abstract group *concept*; dawn of abstract group *theory*

The abstract group concept spread rapidly during the 1880s and 1890s, although there still appeared a great many papers in the areas of permutation and transformation groups. The abstract viewpoint was manifested in two ways:

(a) Concepts and results introduced and proved in the setting of "concrete" groups were now reformulated and reproved in an abstract setting;
(b) Studies originating in, and based on, an abstract setting began to appear.

An interesting example of the former case is the reproving by Frobenius, in an abstract setting, of *Sylow's theorem*, which was proved by Sylow in 1872 for permutation groups. This was done in 1887, in a paper entitled "A new proof of Sylow's theorem." Although Frobenius admitted that the fact that every finite group can be represented by a group of permutations proves that Sylow's theorem must hold for all finite groups, he nevertheless wished to establish the theorem abstractly:

> Since the symmetric group, which is introduced into all these proofs, is totally alien to the context of Sylow's theorem, I have tried to find a new derivation of it.

For a case study of the evolution of abstraction in group theory in connection with Sylow's theorem see [28] and [32].

Hölder was an important contributor to abstract group theory, and was responsible for introducing a number of group-theoretic concepts abstractly. For example, in 1889 he defined the abstract notion of a *quotient group*. The quotient group was first seen as the group of the "auxiliary equation," later as a homomorphic image, and only in Hölder's time as a group of cosets. He then "completed" the proof of the *Jordan–Hölder theorem*, namely that the quotient groups in a composition series are invariant up to isomorphism (see Jordan's contribution, p. 25). For a history of the concept of quotient group see [36].

In 1893, in a paper on groups of order p^3, pq^2, pqr, and p^4, Hölder introduced the concept of an *automorphism* of a group abstractly. He was also the first to study *simple groups* abstractly. (Previously they were considered in concrete cases—as permutation groups, transformation groups, and so on.) As he said: "It would be of the greatest interest if a survey of all simple groups with a finite number of operations could be known." (By "operations" Hölder meant "elements.") He then went on to determine the simple groups of order up to 200.

Other typical examples of studies in an abstract setting are the papers by Dedekind and G. A. Miller in 1897-1898 on Hamiltonian groups, nonabelian groups in which all subgroups are normal. They (independently) characterized such groups abstractly, and introduced the notions of the *commutator* of two elements and the commutator subgroup (Jordan had previously introduced the notion of commutator of two permutations). See [24], [33].

The theory of group characters and the *representation theory* of finite groups, created at the end of the nineteenth century by Burnside, Frobenius, and Molien, also belong to the area of abstract group theory, as they were used to prove important results about abstract groups. See [17] for details.

Although the abstract group *concept* was well established by the end of the nineteenth century, "this was not accompanied by a general acceptance of the associated method of presentation in papers, textbooks, monographs, and lectures.

Group-theoretic monographs based on the abstract group concept did not appear until the beginning of the twentieth century. Their appearance marked the birth of abstract group *theory*" [33].

The earliest monograph devoted entirely to abstract groups was the book by J. A. de Séguier of 1904 entitled *Elements of the Theory of Abstract Groups* [27]. At the very beginning of the book there is a set-theoretic introduction based on the work of Cantor: "De Séguier may have been the first algebraist to take note of Cantor's discovery of uncountable cardinalities" [7]. Next is the introduction of the concept of a semigroup with two-sided cancellation law and a proof that a finite semigroup is a group. There is also a proof of the independence of the group postulates.

De Séguier's book also included a discussion of isomorphisms, homomorphisms, automorphisms, decomposition of groups into direct products, the Jordan–Hölder theorem, the first isomorphism theorem, abelian groups including the basis theorem, Hamiltonian groups, and the theory of p-groups. All this was done in the abstract, with "concrete" groups relegated to an appendix:

> The style of de Séguier is in sharp contrast to that of Dyck. There are no intuitive considerations ... and there is a tendency to be as abstract and as general as possible ... [7].

De Séguier's book was devoted largely to finite groups. The first abstract monograph on group theory which dealt with groups in general, relegating finite groups to special chapters, was O. J. Schmidt's *Abstract Theory of Groups* of 1916 [26]. Schmidt, founder of the Russian school of group theory, devoted the first four chapters of his book to group properties common to finite and infinite groups. Discussion of finite groups was postponed to Chapter 5, there being ten chapters in all. See [7], [10], [33].

2.5 Divergence of developments in group theory

Group theory evolved from several different sources, giving rise to various concrete theories. These theories developed independently, some for over one hundred years, beginning in 1770, before they converged in the early 1880s within the abstract group concept. Abstract group theory emerged and was consolidated in the next thirty to forty years. At the end of that period (around 1920) one can discern the divergence of group theory into several distinct "theories." Here is the barest indication of some of these advances and new directions in group theory, beginning in the 1920s, with the names of some of the major contributors and approximate dates:

(a) Finite group theory. The major problem here, already formulated by Cayley in the 1870s and studied by Jordan and Hölder, was to find all finite groups of a given order. The problem proved too difficult and mathematicians turned to special cases, suggested especially by Galois theory: to find all simple or all solvable groups (cf. the Feit–Thompson theorem of 1963, and the classification of all finite simple groups in 1981). See [14], [15], [30].

(b) Extensions of certain results from finite groups to infinite groups with finiteness conditions. An example is O. J. Schmidt's proof in 1928 of the Remak–Krull–Schmidt theorem. See [5].

(c) Group presentations (combinatorial group theory). This was begun by von Dyck in 1882, and continued in the 20th century by Dehn, Tietze, Nielsen, Artin, Schreier, and others. For a full account see [7].

(d) Infinite abelian group theory. Important in this connection are the works of Prüfer, Baer, Ulm, and others in the 1920s and 1930s. See [30].

(e) Schreier's theory of group extensions (1926). This led to the introduction of the cohomology of groups.

(f) Algebraic groups. Here the work of Borel and Chevalley of the 1940s stands out.

(g) Topological groups, including the extension of group representation theory to continuous groups. Prominent names are Schreier, E. Cartan, Pontrjagin, Gelfand, and von Neumann (1920s and 1930s). See [4].

The figure on the next page gives a diagrammatic sketch of the evolution of group theory as outlined in the various sections and as summarized at the beginning of this section.

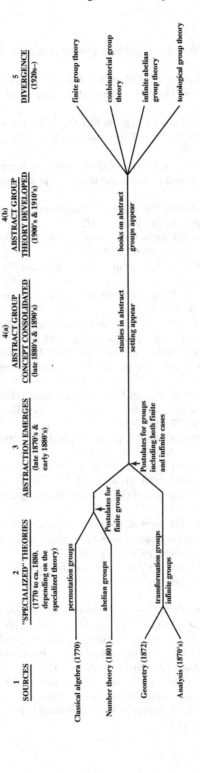

References

1. R. G. Ayoub, Paolo Ruffini's contributions to the quintic, *Arch. Hist. Ex. Sc.* 1980, **23**: 253–277.
2. E. T. Bell, *The Development of Mathematics*, McGraw Hill, 1945.
3. G. Birkhoff, Current trends in algebra, *Amer. Math. Monthly* 1973, **80**: 760–782 and 1974, **81**: 746.
4. G. Birkhoff, The rise of modern algebra to 1936, in *Men and Institutions in American Mathematics*, ed. by D. Tarwater, I. T. White, and I. D. Miller, Texas Tech. Press, 1976, pp. 41–63.
5. N. Bourbaki, *Elements of the History of Mathematics*, Springer-Verlag, 1994.
6. J. E. Burns, The foundation period in the history of group theory, *Amer. Math. Monthly* 1913, **20**: 141–148.
7. B. Chandler and W. Magnus, *The History of Combinatorial Group Theory: A Case Study in the History of Ideas*, Springer-Verlag, 1982.
8. A. Dahan, Les travaux de Cauchy sur les substitutions. Etude de son approche du concept de groupe, *Arch. Hist. Ex. Sc.* 1980, **23**: 279–319.
9. J. Dieudonné (ed.), *Abrégé d'Histoire des Mathématiques, 1700—1900*, 2 vols., Hermann, 1978.
10. P. Dubreil, L'algèbre, en France, de 1900 a 1935, *Cahiers du seminaire d'histoire des mathématiques* 1981, **3**: 69–81.
11. C. H. Edwards, *The Historical Development of the Calculus*, Springer-Verlag, 1979.
12. H. M. Edwards, *Galois Theory*, Springer-Verlag, 1984.
13. A. Gallian, The search for finite simple groups, *Math. Magazine* 1976, **49**: 163–179.
14. D. Gorenstein, *Finite Simple Groups: An Introduction to their Classification*, Plenum Press, 1982.
15. D. Gorenstein, *The Classification of Finite Simple Groups*, Plenum Press, 1983.
16. R. R. Hamburg, The theory of equations in the 18th century: The work of Joseph Lagrange, *Arch. Hist. Ex. Sc.* 1976/77, **16**: 17–36.
17. T. Hawkins, Hypercomplex numbers, Lie groups, and the creation of group representation theory, *Arch. Hist. Ex. Sc.* 1971/72, **8**: 243–287.
18. T. Hawkins, The *Erlanger Programm* of Felix Klein: Reflections on its place in the history of mathematics, *Hist. Math.* 1984, **11**: 442–470.
19. B. M. Kiernan, The development of Galois theory from Lagrange to Artin, *Arch. Hist. Ex. Sc.* 1971/72, **8**: 40–154.
20. F. Klein, *Development of Mathematics in the 19th Century* (transl. from the 1928 German ed. by M. Ackerman), in *Lie Groups: History, Frontiers and Applications*, vol. IX, ed. by R. Hermann, Math. Sci. Press, 1979, pp. 1–361.
21. M. Kline, *Mathematical Thought from Ancient to Modern Times*, Oxford Univ. Press, 1972.
22. D. R. Lichtenberg, *The Emergence of Structure in Algebra*, Doctoral Dissertation, Univ. of Wisconsin, 1966.
23. U. Merzbach, *Quantity to Structure: Development of Modern Algebraic Concepts from Leibniz to Dedekind*, Doctoral Dissertation, Harvard Univ., 1964.
24. G. A. Miller, History of the Theory of Groups, *Collected Works*, 3 vols., pp. 427–467, pp. 1–18, and pp. 1–15, respectively, Univ. of Illinois Press, 1935, 1938, and 1946.
25. L. Novy, *Origins of Modern Algebra*, Noordhoff, 1973.
26. O. J. Schmidt, *Abstract Theory of Groups*, W. H. Freeman & Co., 1966. (Translation by F. Holling and I. B. Roberts of the 1916 Russian edition.)
27. J.-A. de Séguier, *Théorie des Groupes Finis. Elements de la Théorie des Groupes Abstraits*, Gauthier-Villars, Paris, 1904.

28. L. A. Shemetkov, Two directions in the development of the theory of non-simple finite groups, *Russ. Math. Surv.* 1975, **30**: 185–206.

29. R. Silvestri, Simple groups of finite order in the nineteenth century, *Arch. Hist. Ex. Sc.* 1979, **20**: 313–356.

30. J. Tarwater, J. T. White, C. Hall, and M. E. Moore (eds.), *American Mathematical Heritage: Algebra and Applied Mathematics*, Texas Tech. Press, 1981. Has articles by Feit, Fuchs, and MacLane on the history of finite groups, abelian groups, and abstract algebra, respectively.

31. B. L. Van der Waerden, Die Algebra seit Galois, *Jahresbericht der Deutsch. Math. Ver.* 1966, **68**: 155–165.

32. W. C. Waterhouse, The early proofs of Sylow's theorem, *Arch. Hist. Ex. Sc.* 1979/80, **21**: 279–290.

33. H. Wussing, *The Genesis of the Abstract Group Concept*, M.I.T. Press, 1984. (Translation by A. Shenitzer of the 1969 German edition.)

34. A. Cayley, The theory of groups, *Amer. Jour. Math.* 1878, **1**: 50–52.

35. P. M. Neumann, What groups were: a study of the development of the axiomatics of group theory, *Bull. Austral. Math. Soc.* 1999, **60**: 285–301.

36. J. Nicholson, The development and understanding of the concept of quotient group, *Hist. Math.* 1993, **20**: 68–88.

3

History of Ring Theory

Algebra textbooks usually give the definition of a ring first and follow it with examples. Of course the examples came first, and the abstract definition later—much later. So we begin with examples.

Among the most important examples of rings are the integers, polynomials, and matrices. "Simple" extensions of these examples are at the roots of ring theory. Specifically, we have in mind the following three examples:

(a) The integers Z can be thought of as the appropriate subdomain of the field Q of rationals in which to do number theory. (The rationals themselves are unsuitable for that purpose: every rational is divisible by every other (nonzero) rational.) Take a simple extension field $Q(\alpha)$ of the rationals, where α is an algebraic number, that is, a root of a polynomial with integer coefficients. $Q(\alpha)$ is called an algebraic number field; it consists of polynomials in α with rational coefficients. For example, $Q(\sqrt{3}) = \{a + b\sqrt{3} : a, b \in Q\}$.

The appropriate subdomain of $Q(\alpha)$ in which to do number theory—the "integers" of $Q(\alpha)$—consists of those elements that are roots of *monic* polynomials with integer coefficients, polynomials $p(x)$ in which the coefficient of the highest power of x is 1. For example, the integers of $Q(\sqrt{3})$ are $\{a + b\sqrt{3} : a, b \in Z\}$ (this is not obvious). This is our first example.

(b) The polynomial rings $\mathbf{R}[x]$ and $\mathbf{R}[x, y]$ in one and in two variables, respectively, share important properties but also differ in significant ways ($\mathbf{R}$ denotes the real numbers). In particular, while the roots of a polynomial in one variable constitute a discrete set of real numbers, the roots of a polynomial in two variables constitute a curve in the plane—a so-called algebraic curve. Our second example, then, is the ring of polynomials in two (or more) variables.

(c) Square $m \times m$ matrices (for example, over the reals) can be viewed as m^2-tuples of real numbers with coordinate-wise addition and appropriate multiplication obeying the axioms of a ring. Our third example consists, more generally, of n-tuples $\mathbf{R}^n$ of real numbers with coordinate-wise addition and appropriate multiplication, so that the resulting system is a (not necessarily commutative) ring. Such systems, often extensions of the complex numbers, were called in the nineteenth and early twentieth centuries *hypercomplex number systems*.

In what contexts did these examples arise? What was their importance? The answers will lead us to the genesis of ring theory.

Rings fall into two broad categories: commutative and noncommutative. The abstract theories of these two categories came from distinct sources and developed in different directions. Commutative ring theory originated in algebraic number theory, algebraic geometry, and invariant theory. Central to the development of these subjects were the rings of integers in algebraic number fields and algebraic function fields, and the rings of polynomials in two or more variables.

Noncommutative ring theory began with attempts to extend the complex numbers to various hypercomplex number systems. The genesis of the theories of commutative and noncommutative rings dates back to the early nineteenth century, while their maturity was achieved only in the third decade of the twentieth century.

The following is a diagrammatic sketch of the above remarks.

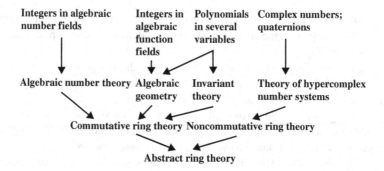

We begin our account with the "simpler" theory of noncommutative rings.

3.1 Noncommutative ring theory

In a strict sense, noncommutative ring theory originated from a single example—the quaternions, invented (discovered?) by Hamilton in 1843. These are "numbers" of the form $a + bi + cj + dk$ (a, b, c, d real numbers) which are added componentwise and in which multiplication is subject to the relations $i^2 = j^2 = k^2 = ijk = -1$. This was the first example of a noncommutative number system, obeying all the (algebraic) laws of the real and complex numbers except for commutativity of multiplication. Such a system is now called a *skew field* or a *division ring* (also a *division algebra*). Hamilton's motivation in introducing the quaternions was to extend the algebra of vectors in the plane to an algebra of vectors in 3-space. Having failed in this task, he turned successfully to quadruples of reals. The "pure" quaternions did, in fact, yield a vector algebra in 3-space. See [19] and Chapter 8.5 for details.

3.1.1 *Examples of Hypercomplex Number Systems*

Hamilton's invention of the quaternions was conceptually groundbreaking—"a revolution in arithmetic which is entirely similar to the one which Lobachevsky effected in geometry," according to Poincaré. Indeed, both achievements were radical violations of prevailing conceptions. Like all revolutions, however, the invention of the quaternions was initially received with less than universal approbation: "I have not yet any clear view as to the extent to which we are at liberty arbitrarily to create imaginaries, and to endow them with supernatural properties," declared Hamilton's mathematician friend John Graves.

Most mathematicians, however, including Graves, soon came around to Hamilton's point of view. The quaternions acted as a catalyst for the exploration of diverse "number systems," with properties which departed in various ways from those of the real and complex numbers. Among the examples of such hypercomplex number systems are the following:

(i) Octonions

These are 8-tuples of reals which contain the quaternions and form a division algebra in which multiplication is nonassociative. They were introduced in 1844 by Cayley and independently by the very John Graves who questioned Hamilton's "imaginaries."

(ii) Exterior algebras

These are n-tuples of reals, added componentwise and multiplied via the "exterior product." They were introduced by Grassmann in 1844 as part of a brilliant attempt to construct a vector algebra in n-dimensional space. Grassmann's style was far from simple and his approach was ahead of its time.

(iii) Group algebras

In 1854 Cayley published a paper on (finite) abstract groups, at the end of which he gave a definition of a group algebra (over the real or complex numbers). He called it a system of "complex quantities" and observed that it is analogous to Hamilton's quaternions—it is associative, noncommutative, but in general not a division algebra.

(iv) Matrices

In two papers of 1855 and 1858 Cayley introduced square matrices. He noted that they can be treated as "single quantities," added and multiplied like "ordinary algebraic quantities," but that "as regards their multiplication, there is the peculiarity that matrices are not in general convertible [commutative]." See Chapter 5.1.3.

(v) Biquaternions

These were introduced by Clifford in 1873 in connection with problems in geometry and physics. They are elements of the form $h_1 + h_2\alpha$, where h_1 and h_2 are quaternions, $\alpha^2 = 1$ and $\alpha h_i = h_i\alpha$.

See [2], [14], [16], [19] for further details.

3.1.2 Classification

Over a thirty-year period (c. 1840–1870) a stock of examples of noncommutative number systems had been established. One could now begin to construct a theory. The general concept of a hypercomplex number system (in current terminology, a finite-dimensional algebra) emerged, and work began on classifying certain types of these structures. We focus on three such developments, dealing with *associative* algebras. Note that such algebras are, of course, rings.

(i) Low-dimensional algebras

Of fundamental importance here was the work of Benjamin Peirce of Harvard—the first important contribution to algebra in the U.S. We are referring to his groundbreaking paper "Linear Associative Algebra" of 1870. In the last 100 pages of this 150-page paper Peirce classified algebras (i.e., hypercomplex number systems) of dimension < 6 by giving their multiplication tables. There are, he showed, over 150 such algebras! What is important in this paper, though, is not the classification but the means used to obtain it. For here Peirce introduced concepts, and derived results, which proved fundamental for subsequent developments. Among these conceptual advances were:

Benjamin Peirce (1809–1880)

(a) An "abstract" definition of a finite-dimensional algebra. Peirce defined such an algebra—he called it a "linear associative algebra"—as the totality of formal expressions of the form $\sum_{i=1}^{n} a_i e_i$, where the e_i are "basis elements." Addition was defined componentwise and multiplication by means of "structural constants" c_{ijk}, namely $e_i e_j = \sum_{k=1}^{n} c_{ijk} e_k$. Associativity under multiplication and distributivity were postulated, but not commutativity. This is probably the earliest explicit definition of an *associative algebra*.

(b) The use of complex coefficients. Peirce took the coefficients a_i in the expressions $\sum a_i e_i$ to be complex numbers. This conscious broadening of the field of coefficients from **R** to **C** was an important conceptual advance on the road to coefficients taken from an arbitrary field.

(c) Relaxation of the requirement that an algebra have an identity. This, too, was a departure from past practice and gave an indication of Peirce's general, abstract approach.

(d) Introduction of nilpotent and idempotent elements. An element x of an algebra is *nilpotent* if $x^n = 0$ for some positive integer n and it is *idempotent* if $x^2 = x$. These concepts proved basic for the subsequent study of algebras and, still later, of rings. Peirce proved the fundamental result that any algebra contains a nilpotent or an idempotent element.

(e) The Peirce decomposition. He showed that if e is an idempotent of an algebra A then $A = eAe \oplus eB_1 \oplus B_2e \oplus B$, where $B_1 = \{x \in A : xe = 0\}$, $B_2 = \{x \in A : ex = 0\}$, and $B = B_1 \cap B_2$ ($\oplus$ indicates direct sum). This so-called *Peirce decomposition* of an algebra relative to an idempotent enabled him to get a better hold on the algebra by studying its constituent parts. It is a central tool in the study of rings and algebras.

Peirce's work was well ahead of its time, and at first attracted little attention. Cayley, for example, who praised it in an address in 1883 to the British Association for the Advancement of Science, called it "outside of ordinary mathematics." Even some of Peirce's admirers in the United States characterized the work as "philosophy of mathematics" rather than mathematics proper. Peirce, of course, turned out to have been a *mathematical* pioneer.

(ii) Division algebras
As we mentioned, the first example of a noncommutative algebra, namely Hamilton's quaternions, was a division algebra. The question arose as to which other finite-dimensional algebras over **R** (algebras of n-tuples of real numbers) are division algebras. The answer was given, independently, by Frobenius (in 1878) and by C.S. Peirce (B. Peirce's son, in 1881): the only such algebras are the real numbers, the complex numbers, and the quaternions.

(iii) Commutative algebras
In the 1860s Dedekind and Weierstrass proved that the only finite-dimensional commutative algebras over **R** or **C**, without nilpotent elements, are direct sums of copies of **R** or **C**. This means that not only addition but also multiplication in such algebras is given componentwise. This result was published only in the 1880s.

See [2], [16], [19] for further details.

3.1.3 Structure

The first example of a noncommutative algebra was given by Hamilton in 1843. During the next forty years mathematicians introduced other examples, began to bring some order into them, and to single out certain types for special attention. The stage was (almost) set for the founding of a general theory of finite-dimensional, noncommutative, associative algebras. The task was accomplished in the last decade of the nineteenth century and the first decade of the twentieth. Before that, however, important developments took place in a neighboring branch of mathematics which had an impact on the work in associative algebras. This was the founding of the theory of Lie groups and Lie algebras in the 1870s and 1880s.

Lie founded the theory of continuous transformation groups (now called Lie groups) in the 1870s to facilitate the study of differential equations (cf. Galois theory). Just as Galois associated a finite (discrete) group of permutations with an algebraic (polynomial) equation, so Lie associated an infinite (continuous) group of transformations with a differential equation. He subsequently showed that for the purposes of the differential equation it suffices to focus on the "local" structure of the Lie group—that is, on the "infinitesimal transformations" which, when multiplied using the "Lie product," form a Lie algebra. (If S, T are infinitesimal transformations, so is their Lie product $[S, T]$, given by $[S, T] = ST - TS$.)

Just as in the case of algebraic equations, so too in this theory, the objects of special interest are the "simple" Lie groups. These give rise to "simple" Lie algebras (i.e., those without ideals). Lie thus proposed the task of studying the structure of Lie algebras with special attention to be given to the "simple" ones. The task was admirably accomplished in the 1880s by Killing and Cartan, who decomposed "semi-simple" Lie algebras (i.e., algebras with zero radical) into simple ones and then classified the latter. See [2], [19].

(i) Algebras over R or C

In the 1890s Cartan, Frobenius, and Molien proved (independently) the following fundamental structure theorem for finite-dimensional associative algebras over the real or complex numbers. If A is such an algebra then

(a) $A = N \oplus B$, where N is nilpotent and B is semi-simple. An algebra N is *nilpotent* if $N^k = 0$ for some positive integer k; it is *semi-simple* if it has no nontrivial nilpotent ideals—this, at least, was the initial conception of semi-simplicity.

(b) $B = C_1 \oplus C_2 \oplus \cdots \oplus C_n$, where C_i are simple algebras, that is, have no nontrivial ideals. (The nilpotent part N is intractable, even today.)

(c) $C_i = M_{n_i}(D_i)$, the algebra of $n_i \times n_i$ matrices with entries from a division algebra D_i.

The above representations are, moreover, unique; that is, the n and n_i are unique, and the N, B, C_i, D_i are unique up to isomorphism.

The immediate inspiration and motivation for this result came from the neighboring theory of Lie algebras (see above). But there were other precedents for decomposition results in algebra—for example, the decomposition of an ideal in the ring of integers of an algebraic number field into a unique product of prime ideals, given by Dedekind in 1871 (see p. 51), and the decomposition of a finite abelian group into a unique direct product of cyclic groups of prime-power order, proved by Frobenius and Stickelberger in 1879 (see p. 28).

Of the work of the three mathematicians who established the above results, Cartan's proved the most influential. His proof techniques, however, were soon superseded by Wedderburn's (see below). What proved lasting, apart from the structure theorem, were the following four concepts which Cartan introduced, albeit only at the *end* of his paper, and only to *state* the structure theorems more succinctly: *direct sum, ideal, simple algebra*, and *semisimple algebra*.

Cartan was the first to introduce these notions explicitly in the context of noncommutative, associative algebras. (Dedekind introduced ideals for certain commutative rings more than two decades earlier, as we shall see below, but there is no reference in Cartan's work to Dedekind's ideals.) For example, he defined an ideal— he called it an "invariant system"—as follows:

> We say that a system $\sum$ admits an invariant subsystem σ, if every element of σ belongs to $\sum$ and if the product, on the right or on the left, of an arbitrary element of $\sum$ and an arbitrary element of σ belongs to σ.

See [16], [19] for further details.

(ii) Algebras over arbitrary fields

At the end of the nineteenth century the theory of finite-dimensional algebras had attained a degree of maturity. All-important connections had been made with Lie's theory of continuous groups as well as with the theory of finite groups, via group representation theory. At the same time, a major structure theorem for associative algebras was available. The theory of finite-dimensional algebras thus became a distinct discipline for serious mathematical investigation. What was needed for further progress in the subject was a new departure. This was provided by Wedderburn's groundbreaking paper of 1907, entitled "On hypercomplex numbers" [20].

The major result in Wedderburn's paper, namely the structure theorem for finite-dimensional algebras, was essentially the same as that given by Cartan. There was "merely" an extension of the field of scalars of the algebra from **R** or **C** to an arbitrary field. This extension, however, necessitated a new approach to the subject—a rethinking and reformulation of its major concepts and results.

Cartan's methods in dealing with the structure of associative algebras over **R** or **C** relied heavily on the vector-space structure of the algebra and on the field of scalars. He associated a characteristic and minimal polynomial with each algebra—these were fundamental tools in his development of the theory. Their factors were related to the structure of the given algebra. For example, he defined a "pseudo-null" element of an algebra as one whose characteristic polynomial has only the zero root. It can be shown that this notion is equivalent to that of a nilpotent element, defined almost thirty years earlier by Benjamin Peirce. See [16].

Wedderburn's approach to the study of the structure of finite-dimensional algebras, which are important examples of noncommutative rings, was conceptual rather than computational. "It is remarkable," he wrote toward the end of the paper, "that the properties of a field with regard to division are not used in many of the theorems of the preceding sections." Among the ideas which he either introduced for the first time or made central in the study of algebras, ideas now—a century later—still recognized by students of algebra as basic, are the notions of *ideal, quotient algebra, nilpotent algebra, radical, semisimple* and *simple algebra, direct sum,* and *tensor product.* His work served as a model for other ring-theoretic structure theorems. See [16], [20].

3.2 Commutative ring theory

Commutative ring theory originated in algebraic number theory, algebraic geometry, and invariant theory, and has in turn been applied mainly to these subjects.

3.2.1 Algebraic Number Theory

Several of the central areas of number theory, principally Fermat's Last Theorem, reciprocity laws, and binary quadratic forms, were instrumental in the emergence of algebraic number theory. Although the main problems in these areas were expressed in terms of integers, it gradually became apparent that the solutions called for embedding the integers in domains of what came to be known as algebraic integers.

(i) Fermat's Last Theorem

Euler in the eighteenth century and several mathematicians in the early nineteenth century realized that to prove Fermat's Last Theorem (FLT)—the unsolvability in nonzero integers of $x^n + y^n = z^n (n > 2)$—even for small values of n, it is necessary to use "complex integers." For example, $x^3 + y^3 = z^3$ is written as $(x + y)(x + y\rho)(x + y\rho^2) = z^3$, where $\rho = (-1 + \sqrt{3}i)/2$ is a primitive cube root of 1, and this is now an equation in the domain $D_3 = \{a + b\rho : a, b \in Z\}$.

Assuming the solvability of $x^3 + y^3 = z^3, z > 0$, and given that D_3 is a unique factorization domain (UFD), one can arrive at a contradiction by showing the existence of integers a, b, c such that $a^3 + b^3 = c^3$, where $0 < c < z$. Repeating the process ad infinitum leads to an infinite descending sequence of positive integers—a contradiction.

If we similarly write $x^p + y^p = z^p$ as $(x+y)(x+y\omega)(x+y\omega^2) \cdots (x+y\omega^{p-1}) = z^p$, where ω is a primitive p-th root of 1 (that is, ω is a root of the equation $x^p = 1$, $\omega \neq 1$), we get an equation in the domain $D_p = \{a_0 + a_1\omega + \cdots + a_{p-1}\omega^{p-1} : a_i \in Z\}$ of so-called *cyclotomic integers*. (Note that it suffices to prove FLT for $n = p$, a prime.) Assuming that D_p is a UFD, we get a contradiction, as above, by showing that there exist integers u, v, w with $u^p + v^p = w^p$ and $0 < w < z$. This, then, proves FLT.

This approach was used by Lamé in 1847, when he announced before the Paris Academy of Sciences that he had proved FLT. The problem with Lamé's proof was his implicit assumption that D_p is a UFD for all primes p. This is of course false, as Kummer pointed out in response to Lamé's purported proof. He showed that D_{23} is not a UFD. It was shown in 1971 that D_p fails to be a UFD for all $p \geq 23$. But viewing $x^p + y^p = z^p$ as an equation in D_p is nevertheless an important idea in dealing with FLT. See [9], [13].

An elementary example of the utility of "complex integers" in solving problems about ordinary integers is the problem of finding all integer solutions of the diophantine equation $x^2 + 2 = y^3$, a special case of the famous "Bachet equation" $x^2 + k = y^3$. It is easy to see that $x = \pm 5$, $y = 3$ are solutions of $x^2 + 2 = y^3$. But are there others?

To find *all* solutions we write $x^2 + 2 = y^3$ as $(x + \sqrt{2}i)(x - \sqrt{2}i) = y^3$. This is now an equation in the domain $D = \{a + b\sqrt{2}i : a, b \in Z\}$. We can show that D is a UFD, and that $x + \sqrt{2}i$ and $x - \sqrt{2}i$ are relatively prime in D. Since their product is a cube, each factor must be a cube (in D). In particular, $x + \sqrt{2}i = (a + b\sqrt{2}i)^3$, for some integers a and b. Cubing and equating coefficients of the real and imaginary parts, we can easily show that $x = \pm 5$, $y = 3$ are the *only* solutions of $x^2 + 2 = y^3$— no easy feat to accomplish without the use of complex integers. This is how Euler (in the eighteenth century) solved the equation.

(ii) Reciprocity laws

Just as solving polynomial equations is important in algebra, solving polynomial congruences, notably $a_0 + a_1 x + \cdots + a_m x^m \equiv 0$ (mod n), is important in number theory. The case of arbitrary m is intractable, but the quadratic case, $a_0 + a_1 x + a_2 x_2 \equiv 0$ (mod n), was dealt with by Gauss in the *Disquisitiones Arithmeticae*. It suffices to consider the congruence $x^2 \equiv q$(mod p), for p and q odd primes (the case of even primes has to be considered separately). Gauss proved the celebrated *quadratic reciprocity law*, namely that $x^2 \equiv q($ mod $p)$ is solvable if and only if $x^2 \equiv p$ (mod q) is solvable, unless $p \equiv q \equiv 3$ (mod 4), in which case $x^2 \equiv q$ (mod p) is solvable if and only if $x^2 \equiv p$ (mod q) is not.

What about higher reciprocity laws? That is, is there a "reciprocity relation" between the solvability of $x^m \equiv q$ (mod p) and $x^m \equiv p$ (mod q) for $m > 2$? Gauss took the view that such laws cannot even be properly conjectured within the context of natural numbers:

> The previously accepted laws of arithmetic are not sufficient for the foundations of a general theory [of reciprocity].... Such a theory demands that the domain of higher arithmetic be endlessly enlarged.

A prophetic statement, indeed. Gauss was calling for the founding of an arithmetic theory of algebraic numbers. In fact, Gauss himself began to enlarge the domain of arithmetic by introducing what came to be known as the *Gaussian integers*, $Z[i] = \{a + bi : a, b \in Z\}$, and showing that they form a UFD. This he did in two papers in 1829 and 1831, in which he used $Z[i]$ to *formulate* the law of biquadratic reciprocity.

At about the same time, Jacobi and Eisenstein (as well as Gauss in unpublished papers) formulated the *cubic reciprocity law*. Here one needed to consider the domain $Z[\rho] = \{a + b\rho : a, b \in Z\}$, where ρ is a primitive cube root of 1 (i.e., $\rho^3 = 1$, $\rho \neq 1$). This was also shown to be a UFD. The search was on for higher reciprocity laws. But as in the case of Fermat's Last Theorem, here too one needed new methods to deal with cases beyond the first few, for unlike $Z[i]$ and $Z[\rho]$, other domains of higher arithmetic needed to formulate such laws were not UFDs. See [4], [13].

(iii) Binary quadratic forms

An (integral) binary quadratic form is an expression of the form $f(x, y) = ax^2 + bxy + cy^2$, $a, b, c \in Z$. The major problem of the theory of quadratic forms was: given a form f, find all integers m which can be represented by f, that is, for which $f(x, y) = m$. For example, Fermat considered the representation of integers n as sums of two squares, $n = x^2 + y^2$.

Gauss in the *Disquisitiones* developed a comprehensive and beautiful theory of binary quadratic forms. Most important was his definition of the composition of two forms and his proof that the (equivalence classes of) forms with a given discriminant $D = b^2 - 4ac$ form (in modern language) a commutative group under this composition. See Chapter 2.2.

The *idea* behind composition of forms is simple: if forms f and g represent integers m and n, respectively, then their composition $f * g$ should represent the product mn. The *implementation* of this idea is subtle and very difficult to describe.

Attempts to gain conceptual insight into Gauss' theory of composition of forms inspired the efforts of some of the best mathematicians of the time, among them Dirichlet, Kummer, and Dedekind. The key idea here, too, was to extend the domain of higher arithmetic and view the problem in a broader context. Here is perhaps the simplest illustration.

If m_1 and m_2 are sums of two squares, so is $m_1 m_2$. Indeed, if $m_1 = x_1^2 + y_1^2$ and $m_2 = x_2^2 + y_2^2$, then $m_1 m_2 = (x_1 x_2 - y_1 y_2)^2 + (x_1 y_2 + x_2 y_1)^2$. In terms of the composition of quadratic forms this can be expressed as $f(x_1, y_1) * f(x_2, y_2) = f(x_1 x_2 - y_1 y_2, x_1 y_2 + x_2 y_1)$, or $f * f = f$, where $f(x, y) = x^2 + y^2$. But even this "simple" law of composition seems mysterious and ad hoc until one introduces the Gaussian integers, which make it transparent:

$$(x_1^2 + y_1^2)(x_2^2 + y_2^2) = (x_1 + y_1 i)(x_1 - y_1 i)(x_2 + y_2 i)(x_2 - y_2 i)$$
$$= (x_1 + y_1 i)(x_2 + y_2 i)(x_1 - y_1 i)(x_2 - y_2 i)$$
$$= (x_1 + y_1 i)(x_2 + y_2 i)\overline{[(x_1 + y_1 i)(x_2 + y_2 i)]} \quad (\bar{\alpha} \text{ denotes the conjugate of } \alpha)$$
$$= [(x_1 x_2 - y_1 y_2) + (x_1 y_2 + x_2 y_1)i][(x_1 x_2 - y_1 y_2) - (x_1 y_2 + x_2 y_1)i]$$
$$= (x_1 x_2 - y_1 y_2)^2 + (x_1 y_2 + x_2 y_1)^2.$$

In general, $ax^2 + bxy + cy^2 = m$ can be written as $(1/a)[ax + (b + \sqrt{D})(y/2)][ax + (b - \sqrt{D})(y/2)] = m$, where $D = b^2 - 4ac$. We have thus expressed the problem of representation of integers by binary quadratic forms in terms of the domain $R = \{(u + v\sqrt{D})/2 : u, v \in Z, u \equiv v \pmod{2}\}$. Since such domains R (for different D) do not, in general, possess unique factorization, the development of their arithmetic theory became an important goal. See [4], [9].

To summarize: we have seen that in dealing with central problems in number theory, namely Fermat's Last Theorem, reciprocity laws, and binary quadratic forms, it was found important to formulate them as problems in domains of algebraic integers. The study of unique factorization in such domains became the major problem of a newly emerging subject—*algebraic number theory*. Kummer dealt with it by means of ideal numbers, Dedekind by means of ideals, and Kronecker by means of divisors. We consider below the contributions of Kummer and Dedekind.

(iv) Kummer's ideal numbers

We recall that the domains of cyclotomic integers, $D_p = \{a_0 + a_1 \omega + \cdots + a_{p-1} \omega^{p-1} : a_i \in Z\}$, ω a primitive p-th root of 1, were central in the study of Fermat's Last Theorem. They also proved important in the investigation of higher reciprocity laws. Both problems were of great interest to Kummer (the latter apparently more than the former), and to make significant progress it was essential to establish unique factorization (of some type) in the domains D_p. This Kummer accomplished in the 1840s. As he put it in a letter to Liouville, unique factorization in D_p "can be saved by the introduction of a new kind of complex numbers that I have called ideal complex numbers." Kummer's major result was that every element in the domain of cyclotomic integers is a unique product of "ideal primes."

Kummer's theory of ideal numbers was vague and computational. In fact, the central notions of ideal number and ideal prime were only *implicitly* defined in terms

Ernst Eduard Kummer (1810–1893)

of their divisibility properties. Kummer noted that in adopting the implicit definitions he was guided by the idea of a "free radical" in chemistry, a substance whose existence can only be discerned by its effects.

To give the reader a sense of Kummer's theory of ideal numbers, we adduce a standard example, due to Dedekind, of a domain $D = \{a+b\sqrt{5}i : a, b \in Z\}$, in which factorization is not unique. We have, for example, $6 = 2 \times 3 = (1 + \sqrt{5}i)(1\sqrt{5}i)$, and it can be readily shown that $2, 3, 1 \pm \sqrt{5}i$ are prime (indecomposable) in D. To restore unique factorization of $6 \in D$, adjoin the "ideal numbers" $\sqrt{2}, (1+\sqrt{5}i)/\sqrt{2}$, and $(1 - \sqrt{5}i)/\sqrt{2}$. These are, in fact, ideal primes.

We then have: $6 = 2 \times 3 = \sqrt{2} \times \sqrt{2} \times [(1 + \sqrt{5}i)/\sqrt{2}] \times [1 - \sqrt{5}i)/\sqrt{2}]$ and $6 = (1 + \sqrt{5}i)(1 - \sqrt{5}i) = \sqrt{2} \times [(1 + \sqrt{5}i)/\sqrt{2}] \times \sqrt{2} \times [1 - \sqrt{5}i)/\sqrt{2}]$. The decomposition of 6 into ideal primes is now unique. Moreover, the choice of the ideal primes $\sqrt{2}, (1 + \sqrt{5}i)/\sqrt{2}$, and $(1 - \sqrt{5}i)/\sqrt{2}$, which seems ad hoc, will come to seem natural after ideals are introduced. See [7], [13].

(v) Dedekind's ideals

Kummer's ideas were brilliant but difficult and not clearly formulated. The fundamental concepts of ideal number and ideal prime were not intrinsically defined. Moreover, Kummer's decomposition theory applied only to cyclotomic integers. What was needed was a decomposition theory which would apply to more general domains of algebraic integers. This was devised, independently and in different ways, by Dedekind and Kronecker. We will focus on Dedekind's formulation, which is the one that has generally prevailed.

The main result of Dedekind's groundbreaking 1871 work, which appeared as Supplement X to the 2nd edition of Dirichlet's *Lectures on Number Theory* (edited by Dedekind) was that every nonzero ideal in the domain of integers of an algebraic number field is a unique product of prime ideals. Before one could state this theorem one had, of course, to define the concepts in its statement, namely "the domain of integers of an algebraic number field," "ideal," and "prime ideal." It took Dedekind about twenty years to do that.

The number-theoretic domains studied at the time, such as the Gaussian integers, the integers arising from cubic reciprocity, and the cyclotomic integers, were all of the form $Z[\theta] = \{a_0 + a_1\theta + \cdots + a_n\theta^n : a_i \in Z\}$, where θ satisfies a polynomial with integer coefficients. It was therefore tempting to define the domains to which Dedekind's theorem would apply as objects of this type. But Dedekind showed that these were the wrong objects. For example, he showed that Kummer's theory of unique factorization could *not* be extended to the domain $Z[\sqrt{3}i] = \{a + b\sqrt{3}i : a, b \in Z\}$, and, of course, Dedekind's objective was to try to extend Kummer's theory to *all* "domains of integers of algebraic number fields" (see below).

One had to begin the search for the appropriate domains, Dedekind contended, within an *algebraic number field*—a finite field extension $Q(\alpha) = \{q_0 + q_1\alpha + \cdots + q_n\alpha^n : q_i \in Q\}$ of the rationals, where α is an algebraic number. The notion of "algebraic number" was well known at the time, but not that of "algebraic integer." Dedekind showed that *all* elements of $Q(\alpha)$ are algebraic numbers.

But what is the appropriate subdomain of $Q(\alpha)$ in which to do number theory—*the integers of $Q(\alpha)$*? Dedekind defined them to be the elements of $Q(\alpha)$ which are roots of *monic* polynomials with integer coefficients, polynomials with coefficient of the term of highest degree equal 1. (Note that under this definition the "ordinary" integers Z—"the integers of Q"—are the roots of *linear* monic polynomials.) He showed that the integers of $Q(\alpha)$ "behave" like the ordinary integers—they are closed under addition, subtraction, and multiplication; in our terminology, they form a ring—a subring of **C**.

But Dedekind did not motivate his definition of the domain of integers of an algebraic number field, as historian of mathematics Edwards laments: "Insofar as this is the crucial idea of the theory, the genesis of the theory appears, therefore, to be lost" [13].

Having defined the domain of algebraic integers of $Q(\alpha)$ in which he would formulate and prove his result on unique decomposition of ideals, Dedekind considered, more generally, sets of integers of $Q(\alpha)$ closed under addition, subtraction, and multiplication. He called them *orders*. (The domain of "integers of $Q(\alpha)$" is the largest order.) Here, then, was an algebraic first—an essentially axiomatic definition of a (commutative) ring, albeit in a concrete setting.

The second fundamental concept of Dedekind's theory, that of ideal, derived its motivation (and name) from Kummer's ideal numbers. Dedekind wanted to characterize them *internally*, within the domain D_p of cyclotomic integers. Thus, for each ideal number σ he considered the set of cyclotomic integers divisible by σ. These, he noted, are closed under addition and subtraction, as well as under multiplication by all elements of D_p. Conversely, he proved (and this is a difficult theorem) that every set of cyclotomic integers closed under these operations is precisely the set of cyclotomic integers divisible by some ideal number τ.

Thus there is a one-one correspondence between ideal numbers and subsets of the cyclotomic integers closed under the above operations. Such subsets of D_p Dedekind called *ideals*. These subsets, then, characterized ideal numbers internally, and served as motivation for the introduction of ideals in arbitrary domains of algebraic integers. Dedekind defined them abstractly as follows:

A subset I of the integers R of an algebraic number field K is an *ideal* of R if it has the following two properties:
  (i) If $\beta, \gamma \in I$ then $\beta \pm \gamma \in I$.
  (ii) If $\beta \in I, \mu \in R$, then $\beta\mu \in I$.

This procedure was typical of Dedekind's modus operandi: He would distill from a concrete object (in this case the set of cyclotomic integers divisible by an ideal number) the properties of interest to him (in this case closure under the above operations) and proceed to define an abstract object (in this case an ideal) in terms of those properties. This is, of course, standard practice nowadays, but it was revolutionary in Dedekind's time.

Dedekind defined a prime ideal—perhaps the most important notion of commutative algebra—as follows: An ideal P of R is *prime* if its only divisors are R and P. Given ideals A and B, A was said to *divide* B if A contains B. In later versions of his work Dedekind showed that A divides B if and only if $B = AC$ for some ideal C of R. Having defined the notion of prime ideal, he proved his fundamental theorem that every nonzero ideal in the ring of integers of an algebraic number field is a unique product of prime ideals. See [7], [8] for details.

How did Dedekind's ideas apply to (say) the nonunique factorization of 6 into primes in the domain $D = \{a + b\sqrt{5}i : a, b \in Z\}$, where, we recall, $6 = 2 \times 3 = (1 + \sqrt{5}i)(1 - \sqrt{5}i)$ (see p. 51)? If we let $P = \langle 2, 1 + \sqrt{5}i \rangle = \{2\alpha + (1 + \sqrt{5}i)\beta : \alpha, \beta \in D\}$—the ideal of D generated by 2 and $1 + \sqrt{5}i$, $Q = \langle 3, 1 + \sqrt{5}i \rangle$, $R = \langle 3, 1 - \sqrt{5}i \rangle$, we can verify that $P^2 = \langle 2 \rangle$, $PQ = \langle 1 + \sqrt{5}i \rangle$, $QR = \langle 3 \rangle$, and $PR = \langle 1 - \sqrt{5}i \rangle$. ($\langle \alpha \rangle$ denotes the ideal generated by α; if A and B are ideals of a ring, their product is the ideal $AB = \{\sum_{\text{finite}} a_i b_i : a_i \in A, b_i \in B\}$.)

The factorizations $6 = 2 \times 3 = (1 + \sqrt{5}i)(1 - \sqrt{5}i)$ now yield the following factorizations of ideals: $\langle 6 \rangle = \langle 2 \rangle \langle 3 \rangle = P^2(QR)$ and $\langle 6 \rangle = \langle 1 + \sqrt{5}i \rangle \langle 1 - \sqrt{5}i \rangle = (PQ)(PR) = P^2QR$. One can readily verify that the ideals P, Q, R are prime. Thus the *ideal* $\langle 6 \rangle$ (if not the *element* 6) has been factored uniquely into prime ideals. Paradise regained via ideals.

Let us now compare the factorization of $\langle 6 \rangle$ into *prime ideals* with the factorization of 6 into *ideal primes* (à la Kummer) that we gave earlier:

$$6 = 2 \times 3 = \sqrt{2} \times \sqrt{2} \times [(1 + \sqrt{5}i)/\sqrt{2}] \times [1 - \sqrt{5}i)/\sqrt{2}] \quad \text{and}$$

$$6 = (1 + \sqrt{5}i)(1 - \sqrt{5}i) = \sqrt{2} \times [(1 + \sqrt{5}i)/\sqrt{2}] \times \sqrt{2} \times [1 - \sqrt{5}i)/\sqrt{2}].$$

Performing some eighteenths century callisthenics, we obtain the following:

Since $P^2 = \langle 2 \rangle$, $P \sim \sqrt{2}$ (where "$\sim$" stands for "corresponds to," "captures," "represents"). In fact, P is the ideal consisting of all elements of D divisible by the ideal number $\sqrt{2}$, that is, such that the quotient is an algebraic integer.

We also have $PQ = \langle 1 + \sqrt{5}i \rangle$, hence $PQ/P \sim (1 + \sqrt{5}i)/\sqrt{2}$, so that the ideal Q corresponds to the ideal number $(1 + \sqrt{5}i)/\sqrt{2}$. And since $PR = \langle 1 - \sqrt{5}i \rangle$,

$PR/P \sim (1 - \sqrt{5}i)/\sqrt{2}$, hence $R \sim (1 - \sqrt{5}i)/\sqrt{2}$. This removes the mystery associated with our earlier introduction of the ideal numbers $\sqrt{2}$, $(1 + \sqrt{5}i)/\sqrt{2}$, and $(1 - \sqrt{5}i)/\sqrt{2}$.

Dedekind's work was the culmination of seventy years of investigations of problems related to unique factorization. It created, in one swoop, a new subject—algebraic number theory. It introduced, albeit in a concrete setting, some of the most fundamental concepts of commutative algebra, such as ring, ideal, and prime ideal. His work also established one of the central results of algebraic number theory, namely the representation of ideals in domains of integers of algebraic number fields as unique products of prime ideals. The theorem was soon to play a fundamental role in the study of algebraic curves (see below).

As important as his concepts and results were Dedekind's methods. In fact, "his insistence on philosophical principles was responsible for many of his important innovations" [8]. One of his philosophical principles was a focus on intrinsic, conceptual properties over formulas, calculations, or concrete representations. Another was the acceptance of nonconstructive procedures (in definitions and proofs) as legitimate mathematical methods. His great concern for teaching also influenced his mathematical thinking. His two very significant methodological innovations were the use (outside of geometry) of the axiomatic method and the institution of set-theoretic modes of thinking. See Chapter 8.2.

The axiomatic method was just beginning to resurface after 2000 years of near dormancy. Dedekind was instrumental in pointing to its mathematical power and pedagogical value. In this he inspired (among others) David Hilbert and Emmy Noether. His use of set-theoretic formulations (recall, for example, his definition of an ideal as the set of elements of a domain satisfying certain properties), including the use of the completed infinite—taboo at the time—preceded by about ten years Cantor's seminal work on the subject.

3.2.2 Algebraic Geometry

Algebraic geometry is the study of algebraic curves and their generalizations to n dimensions, algebraic varieties. An *algebraic curve* is the set of roots of an algebraic function; that is, a function $y = f(x)$ defined implicitly by the polynomial equation $P(x, y) = 0$. It is natural to study algebraic curves in complex projective space.

Several approaches were used in the study of algebraic curves, notably the analytic, the geometric-algebraic, and the algebraic-arithmetic. In the analytic approach, to which Riemann (in the 1850s) was the major contributor, the main objects of study were algebraic functions $f(w, z) = 0$ of a complex variable and their integrals, the so-called abelian integrals, which are closely related to the important notion of the genus of an algebraic curve. It was in this connection that Riemann introduced the fundamental concept of a Riemann surface, on which algebraic functions become single-valued. Riemann's methods were, however, nonrigorous, relying heavily on the physically obvious, but mathematically questionable, Dirichlet Principle.

In the 1860s and 1870s Clebsch, Gordan, Brill, and especially M. Noether introduced geometric-algebraic methods to study algebraic functions and curves.

A major problem, solved by Noether, was: given algebraic curves $f(x, y) = 0$ and $g(x, y) = 0$, to find conditions under which a polynomial $F(x, y)$ is representable in the form $F = Af + Bg$, where A and B are polynomials in x and y. In modern terms: under what conditions is F an element of the ideal (in the polynomial ring $\mathbf{R}[x, y]$) generated by f and g? The ideas of the geometric-algebraic school can be thought of as the starting point of the theory of polynomial ideals. See [2], [10], [12] for details.

(i) Algebraic function fields

Neither the transcendental methods of Riemann, nor the geometric-algebraic ideas of M. Noether et al., provided a rigorous foundation for algebraic function theory. This was accomplished by Dedekind and Weber in their groundbreaking 1882 paper "Theory of algebraic functions of a single variable," in which they proposed to "provide a basis for the theory of algebraic functions, the major achievement of Riemann's researches, in the simplest and at the same time rigorous and most general manner." The fundamental idea of their algebraic-arithmetic approach was to carry over to algebraic function fields the ideas which Dedekind had earlier introduced for algebraic number fields.

Just as an algebraic number field is a finite extension $Q(\alpha)$ of the field Q of rationals, so an algebraic function field is a finite extension $K = \mathbf{C}(z)(w)$ of the field $\mathbf{C}(z)$ of rational functions (in the indeterminate z). That is, w is a root of a polynomial $a_0 + a_1\alpha + a_2\alpha^2 + \cdots + a_n\alpha^n$, where $a_i \in \mathbf{C}(z)$ (we can take $a_i \in \mathbf{C}[z]$). Thus $w = f(z)$ is an algebraic function defined implicitly by the polynomial equation $P(z, w) = a_0 + a_1w + a_2w^2 + \cdots + a_nw^n = 0$. In fact, *all* the elements of $K = \mathbf{C}(z)(w) = \mathbf{C}(z, w)$ are algebraic functions.

Let now A be the "ring of integers" of K over $\mathbf{C}(z)$; that is, A consists of the elements of K which are roots of *monic* polynomials over $\mathbf{C}[z]$. As for algebraic numbers, here too every nonzero ideal of A is a unique product of prime ideals (cf. p. 52). (Incidentally, the meromorphic functions on a Riemann surface form a field of algebraic functions, with the entire functions as their "ring of integers.")

Dedekind and Weber were now ready to give a rigorous, algebraic definition of a Riemann surface S of the algebraic function field K: it is (in our terminology) the set of nontrivial discrete valuations on K. (The finite points of S correspond to ideals of A; to deal with points at infinity of S Dedekind and Weber introduced the notions of "place" and "divisor.") Many of Riemann's ideas on algebraic functions were here developed algebraically and rigorously. In particular, a rigorous proof was given of the important Riemann–Roch theorem. See [2], [10].

Beyond Dedekind and Weber's technical achievements in putting major parts of Riemann's algebraic function theory on solid ground, their conceptual breakthrough lay in pointing to the strong analogy between algebraic number fields and algebraic function fields, hence between algebraic number theory and algebraic geometry. This analogy proved extremely fruitful for both theories. For example, the use of power series in algebraic geometry inspired Hensel in 1897 to introduce p-adic numbers ("power series" in the prime p; see Chapter 4.7). The resulting idea of p-adic completion proved important in both algebraic number theory and algebraic geometry. Another noteworthy aspect of Dedekind and Weber's work was its generality and applicability to arbitrary fields, in particular Q and Z_p, which were important in number-theoretic contexts. Thus ideas from algebraic geometry could be applied to number theory. See [10], [17].

(ii) Polynomial rings and their ideals

As we noted, polynomial ideals in algebraic geometry had their implicit beginnings in M. Noether's work (c. 1870). Important advances were made by Kronecker in the 1880s and especially by Hilbert, Lasker, and Macauley in 1890, 1905, and 1913, respectively.

The need for polynomial ideals in the study of algebraic varieties is manifest. An *algebraic variety* V is defined as the set of points in $\mathbf{R}^n$ (or $\mathbf{C}^n$) satisfying a system of polynomial equations $f_i(x_1, x_2, \ldots, x_n) = 0, i = 1, 2, \ldots$. The Hilbert basis theorem implies that finitely many equations will do. But different systems of polynomial equations may give rise to the same set of roots. For example, the circle V in $\mathbf{R}^3$ of radius 2 lying in the plane parallel to the x-y plane and two units above it may be described as $V = \{(x, y, z) : x^2 + y^2 - 4 = 0, z - 2 = 0\}$, as $V = \{(x, y, z) : x^2 + y^2 + z^2 - 8 = 0, z - 2 = 0\}$, or as $V = \{(x, y, z) : x^2 + y^2 - 4 = 0, x^2 + y^2 - 2z = 0\}$. Is there a canonical set of polynomials which describes the variety (circle) V?

It is easy to see that if $f_1, f_2, \ldots, f_m$ are polynomials which vanish on the points of V, then so do all polynomials of the set $I = \{g_1 f_1 + g_2 f_2 + \cdots + g_m f_m : g_i \in \mathbf{R}[x, y, z]\}$. But I is an ideal of the polynomial ring $\mathbf{R}[x, y, z]$. In fact, the set of *all* polynomials of $\mathbf{R}[x, y, z]$ which vanish on the points of V is also an ideal—and it is evidently the "canonical" set of polynomials to describe V.

Note that the above remarks point to a correspondence between ideals of $\mathbf{R}[x_1, x_2, \ldots, x_n]$ (or of $\mathbf{C}[x_1, x_2, \ldots, x_n]$) and varieties in $\mathbf{R}^n$ (or $\mathbf{C}^n$): If V is a variety, let $I(V) = \{f(x_1, x_2, \ldots, x_n) \in \mathbf{R}[x_1, x_2, \ldots, x_n] : f(a_1, a_2, \ldots, a_n) = 0$ for all $(a_1, a_2, \ldots, a_n) \in V\}$, and if J is an ideal of $\mathbf{R}[x_1, x_2, \ldots, x_n]$, let $V(J) = \{(b_1, b_2, \ldots, b_n) \in \mathbf{R}^n : g(b_1, b_2, \ldots, b_n) = 0$ for all $g \in J\}$.

The Hilbert Nullstellensatz (in one of its incarnations) says that $V(J) \neq \phi$ if the variety is in $\mathbf{C}^n$ (or K^n for any algebraically closed field K). This correspondence is central in algebraic geometry. It is, in fact, a one-one correspondence between varieties over an algebraically closed field K and their largest defining ideals, the so-called radical ideals. Under this correspondence prime ideals correspond to irreducible varieties. See [5], [10], [12].

Hilbert, Lasker, and Macauley exploited the above correspondence in the latter nineteenth and early twentieth centuries by undertaking a thorough study of ideals in polynomial rings in order to shed light on algebraic varieties. Lasker's major result was the "primary decomposition" of ideals: every ideal in a polynomial ring $F[x_1, x_2, \ldots, x_n]$ is a finite intersection of primary ideals. (Primary ideals, first defined by Lasker, are generalizations of prime ideals; the former are to the latter what prime powers are to primes in the ring of integers.) Translated into the language of algebraic geometry, the result says that every variety is a finite union of irreducible varieties, that is, those that cannot be nontrivially decomposed as finite unions of other varieties. Macauley proved the uniqueness of the primary decomposition, which implied that

David Hilbert (1862–1943)

every variety can be expressed uniquely as a union of irreducible varieties—a type of fundamental theorem of arithmetic for varieties. Hilbert's important contributions to the subject were made in the context of his work on invariants (see below).

See [5], [10], [12] for further details.

3.2.3 Invariant Theory

Invariant theory has its roots in both number theory and geometry. Given two binary quadratic forms $f = ax^2 + bxy + cy^2$ and $f_1 = a_1x_1^2 + b_1x_1y_1 + c_1y_1^2$ over the integers, Gauss defined them to be *equivalent* if f can be changed into f_1 by a linear transformation given by $x = px_1 + qy_1$, $y = rx_1 + sy_1$, where $ps - qr = 1$. Equivalent quadratic forms represent the same set of integers. Moreover, the discriminants $D = b^2 - 4ac$ and $D_1 = b_1^2 - 4a_1c_1$ of f and f_1, respectively, are equal. The discriminant is thus said to be an *invariant* of the quadratic form under a linear transformation of the variables with determinant 1.

The first half of the nineteenth century saw the rise of new geometries—projective, hyperbolic, Riemannian, algebraic, and others. Efforts were undertaken to distinguish among the different types of geometry by pinpointing the characteristics of each. Invariance of properties under various transformations was an important tool in these studies, leading in time to Klein's Erlangen Program. For example, projective properties of geometric figures are those which are invariant under linear transformations, while algebraic-geometric properties are those invariant under birational transformations (see Chapter 2.1.3).

In the mid-nineteenth century invariant theory became an independent field of study, divorced from its number-theoretic and geometric connections. In fact, between the 1860s and the 1880s it became a major branch of *algebra*. Two problems engaged mathematicians' interest: to find specific invariants of various forms and to find "complete systems" of invariants.

Specifically, given a binary form $f_1(x_1, x_2) = a_0 x_1^n + a_1 x_1^{n-1} x_2 + \cdots + a_n x_2^n$ (the a_i now taken in **R** or **C**), let it be changed by a linear transformation of the variables x_1, x_2 into the form $F(X_1, X_2) = A_0 X_1^n + A_1 X_1^{n-1} X_2 + \cdots + A_n X_2^n$. A function I of the coefficients which satisfies the relation $I(A_0, A_1, \ldots, A_n) = r^k I(a_0, a_1, \ldots, a_n)$ (r in **R** or **C**) is called an *invariant* of f (under linear transformations). Cayley, Sylvester, Gordan and others found specific invariants (e.g., the Jacobian, the Hessian) for specific forms (e.g., binary quartic forms, cubic forms).

Attention turned, in time, to finding a *complete system of invariants* for a given form, namely a minimal set of invariants such that any other invariant of the form could be expressed as a linear combination of the minimal set. The existence of a *finite* complete system—a basis—for *binary* forms of any degree was first established by Gordan in 1868. His proof was long and difficult and showed how to *compute* the basis. Bases for a number of other forms (e.g., ternary quadratic and cubic forms) were obtained during the next twenty years. See Chapter 8.1.1.

Hilbert, who wrote a thesis on invariants in 1885, and in 1888 gave a much simpler, but noncomputational, proof of Gordan's result on binary forms, astonished the mathematical community in 1890 by showing that any form, of any degree, in any number of variables, has a basis. Hilbert adopted a new, conceptual, approach to the subject. The idea was to consider, instead of invariants, expressions in a finite number of variables—in short, the polynomial ring in those variables. Hilbert then proved what came to be known as *Hilbert's Basis Theorem*, namely that every ideal in the ring of polynomials in finitely many variables has a finite basis. The existence of a basis for an arbitrary form now followed. "This is not mathematics, it is theology," protested Gordan in response to Hilbert's abstract, nonconstructive proof. The theology of the 1890s, however, became the mathematical gospel of the 1920s. See [6], [10], [14].

3.3 The abstract definition of a ring

In the first decade of the twentieth century there were well-established, flourishing, concrete theories of both commutative and noncommutative rings and their ideals (the noncommutative theory dealt with algebras, which are of course rings). Their roots were mainly in algebraic number theory, algebraic geometry, and the theory of hypercomplex numbers. Moreover, abstract (axiomatic) definitions of groups, fields, and vector spaces had then been in existence for about two decades. The time was ripe for the abstract ring concept to emerge.

The first abstract definition of a ring was given by Fraenkel (of set-theory fame) in a 1914 paper entitled "On zero divisors and the decomposition of rings" [11]. Fraenkel's definition meant to encompass both commutative and noncommutative rings, for the examples of rings he gave included integers modulo n, matrices, p-adic integers, and hypercomplex number systems. But his work was not grounded in the major concrete theories which had earlier been established.

Fraenkel's aim in this paper was to do for rings what Steinitz had just (1910) done for fields (see Chapter 4), namely to give an abstract and comprehensive theory of commutative and noncommutative rings. Of course he was not successful (he did admit that the task here was not as "easy" as in the case of fields)—it was too ambitious an undertaking to subsume the structure of both commutative and noncommutative rings under one theory.

Among the main concepts introduced in Fraenkel's paper are "zero divisors" and "regular elements." Fraenkel considered only rings which are not integral domains (i.e., rings with zero divisors) and discussed divisibility for such rings. Much of the paper dealt with decomposition of rings as direct products of "simple" rings (not the usual notion of simplicity).

Fraenkel's definition of a ring is in today's style. He defined it as "a system" with two abstract operations, to which he gave the names addition and multiplication. Under one of the operations (addition) the system forms a group—he gave its axioms. The second operation (multiplication) is associative and distributes over the first. Two axioms give the closure of the system under the operations, and there is the requirement of an identity in the definition of the ring.

Commutativity under addition did *not* appear as an axiom but was proved! So were other elementary properties of a ring, such as $a \times 0 = 0$, $a(-b) = (-a)b = -(ab)$, and $(-a)(-b) = ab$. There were two "extraneous" axioms, dealing with "regular" elements in the ring, which departed from an otherwise modern definition.

The latter was given by Sono in a 1917 paper entitled "On congruences" [18]. Sono's was a very modern, abstract work, discussing cosets, quotient rings, maximal and minimal ideals, simple rings, the isomorphism theorems, and composition series.

Although Fraenkel's and Sono's works were not in the mainstream of contemporary ring-theoretic studies, their significance was that rings now began to be studied as independent, abstract objects, not just as rings of polynomials, as rings of algebraic integers, or as rings (algebras) of hypercomplex numbers.

3.4 Emmy Noether and Emil Artin

Despite the abstract definition of a ring, rings of polynomials, rings of algebraic integers, and rings of hypercomplex numbers remained central in ring theory. In the hands of the master algebraists Noether and Artin their study was transformed in the 1920s into powerful, abstract theories. Noether's two seminal papers of 1921 and 1927 extended and abstracted the decomposition theories of polynomial rings on the one hand and of the rings of integers of algebraic number fields and algebraic function fields on the other, to abstract commutative rings with the ascending chain condition—now called *Noetherian rings*.

More specifically, Noether showed in the 1921 paper, entitled "Ideal theory in rings," that the results of Hilbert, Lasker, and Macauley on primary decomposition in polynomial rings hold for any (abstract) ring with the ascending chain condition. Thus results which seemed inextricably connected with the properties of polynomial rings were shown to follow from a single axiom!

In her 1927 paper, "Abstract development of ideal theory in algebraic number fields and function fields," she discussed the Dedekind and Dedekind-Weber results on decomposition of ideals as unique products of prime ideals in, respectively, rings of integers of algebraic number fields and function fields, in the setting of abstract rings. In particular, she characterized abstract commutative rings in which every nonzero ideal is a unique product of prime ideals. Such rings are now called *Dedekind domains*. See Chapter 6 for details.

Artin, inspired by Noether's work on commutative rings with the ascending chain condition, generalized Wedderburn's structure theorems on algebras (see p. 47) in his 1927 paper, "On the theory of hypercomplex numbers," to noncommutative rings with the descending chain condition. In particular, he showed that such rings, with zero radical—now called *Artinian rings*, can be decomposed into direct sums of simple rings which, in turn, are matrix rings over division rings.

While with Fraenkel and Sono we witness the birth of the abstract ring *concept*, with Noether and Artin we see the birth of abstract ring *theory*. Noether and Artin made the abstract ring concept central in algebra by framing in an abstract setting the theorems which were its major inspirations. In this context they introduced and gave prominence to such fundamental algebraic notions as ideal (including one-sided ideal), module, and chain conditions—both ascending and descending. Ring theory now took its rightful place along the by then well established theories of groups and fields as one of the pillars of abstract algebra.

See [2], [10], [19] for further details.

3.5 Epilogue

The importance of ring theory in algebra and beyond has anything but diminished in the seventy or so years since Noether's and Artin's works. To illustrate we quote from a 1991 book on the subject, *A First Course in Noncommutative Rings*, by the prominent algebraist T. Y. Lam:

> Today, ring theory is a fertile meeting ground for group theory (group rings), representation theory (modules), functional analysis (operator algebras), Lie theory (enveloping algebras), algebraic geometry (finitely generated algebras, differential operators, invariant theory), arithmetic (orders, Brauer groups), universal algebra (varieties of rings), and homological algebra (cohomology of rings, projective modules, Grothendieck and higher K-groups).

As a final comment, the paper of Richard Taylor and Andrew Wiles, filling a gap in Wiles' previously announced proof of Fermat's Last Theorem, is entitled "Ring-theoretic properties of certain Hecke algebras" (see *Ann. Math.* 1995, **141**: 553–572).

References

1. G. Birkhoff and S. MacLane, *A Survey of Modern Algebra*, Macmillan, 1941.

2. N. Bourbaki, *Elements of the History of Mathematics*, Springer-Verlag, 1994.

3. D. M. Burton and D. H. Van Osdol, Toward the definition of an abstract ring, in *Learn from the Masters*, ed. by F. Swetz et al, Math. Assoc. of America, 1995, pp. 241–251.

4. H. Cohn, *Advanced Number Theory*, Dover, 1980.

5. D. Cox, J. Little, and D. O'Shea, *Ideals, Varieties, and Algorithms*, Springer-Verlag, 1992.

6. T. Crilly, Invariant theory, in *Companion Encyclopedia of the History and Philosophy of the Mathematical Sciences*, ed. by I. Grattan-Guinness, Routledge, 1994, pp. 787–793.

7. H. M. Edwards, Dedekind's invention of ideals, *Bull. Lond. Math. Soc.* 1983, **15**: 8–17.

8. H. M. Edwards, The genesis of ideal theory, *Arch. Hist. Ex. Sci.* 1980, **23**: 321–378.

9. H. M. Edwards, *Fermat's Last Theorem: A Genetic Introduction to Algebraic Number Theory*, Springer-Verlag, 1977.

10. D. Eisenbud, *Commutative Algebra, with a View Toward Algebraic Geometry*, Springer-Verlag, 1995.

11. A. Fraenkel, Über die Teiler der Null und die Zerlegung von Ringen, *Jour. für die Reine und Angew. Math.* 1914, **145**: 139–176.

12. J. J. Gray, Early modern algebraic geometry, in *Companion Encyclopedia of the History and Philosophy of the Mathematical Sciences*, ed. by I. Grattan-Guinness, Routledge, 1994, pp. 920–926.

13. K. Ireland and M. Rosen, *A Classical Introduction to Modern Number Theory*, 2nd ed., Springer-Verlag, 1982.

14. M. Kline, *Mathematical Thought from Ancient to Modern Times*, Oxford University Press, 1972.

15. C. C. MacDuffee, Algebra's debt to Hamilton, *Scripta Math.* 1944, **10**: 25–35.

16. K. H. Parshall, H.M. Wedderburn and the structure theory of algebras, *Arch. Hist. Ex. Sci.* 1985, **32**: 223–349.

17. J. H. Silverman and J. Tate, *Rational Points on Elliptic Curves*, Springer-Verlag, 1992.

18. M. Sono, On congruences, I-IV, *Mem. Coll. Sci. Kyoto* 1917, **2**: 203–226, 1918, **3**: 113–149, 189–197, and 299–308.

19. B. L. van der Waerden, *A History of Algebra*, Springer-Verlag, 1985.

20. J. H. M. Wedderburn, On hypercomplex numbers, *Proc. Lond. Math. Soc.* 1907, **6**: 77–118.

4

History of Field Theory

The evolution of field theory spans a period of about 100 years, beginning in the early decades of the nineteenth century. This period also saw the development of the other major algebraic theories, namely group theory, ring theory, and linear algebra. The evolution of field theory was closely intertwined with that of the other three theories, as we shall see.

Abstract field theory emerged from three concrete theories—what came to be known as Galois theory, algebraic number theory, and algebraic geometry. These were founded, and began to flourish, in the nineteenth century. Of some influence in the rise of the abstract field concept were also the theory of congruences and (British) symbolical algebra. The nineteenth century's increased concern for rigor, generalization, and abstraction undoubtedly also had an impact on our story.

In this chapter we shall discuss the sources of field theory as well as some of the main events in its evolution, culminating in Steinitz's abstract treatment of fields.

4.1 Galois theory

For three millennia, until the early nineteenth century, algebra meant solving polynomial equations, mainly of degrees up to four. Field-theoretic ideas are implicit even here. For example, in solving the linear equation $ax + b = 0$, the four algebraic operations come into play and hence implicitly the notion of a field. In the case of the quadratic equation $ax^2 + bx + c = 0$, its solutions, $x = (-b \pm \sqrt{b^2 - 4ac})/2a$, require the adjunction of square roots to the field of coefficients of the equation. The concept of adjunction of an element to a field is fundamental in field theory.

Field-theoretic notions appear much more prominently, even if at first still implicitly, in the *modern* theory of solvability of polynomial equations. The groundwork was laid by Lagrange in 1770, but the field-theoretic elements of the subject were introduced by Abel and Galois in the early decades of the nineteenth century. Ruffini's 1799 proof of the insolvability of the quintic had a major gap because he lacked sufficient understanding of field-theoretic ideas.

Such ideas were starting points in Galois' 1831 "Memoir on the conditions of solvability of equations by radicals":

> One can agree to regard all rational functions of a certain number of determined quantities *a priori*. For example, one can choose a particular root of a whole number and regard as rational every rational function of this radical. When we agree to regard certain quantities as known in this manner, we shall say that we *adjoin* them to the equation to be resolved. We shall say that these quantities are *adjoined* to the equation. With these conventions, we shall call *rational* any quantity which can be expressed as a rational function of the coefficients of the equation and of a certain number of *adjoined* quantities arbitrarily agreed upon...One can see, moreover, that the properties and the difficulties of an equation can be altogether different, depending on what quantities are adjoined to it.

It is clear that Galois has a good insight into the fields which we would denote today by $F(u_1, u_2, \ldots, u_n)$, obtained by adjoining the quantities $u_1, u_2, \ldots, u_n$ to the (field of) coefficients of an equation. In the specific example mentioned, Galois has in mind a quadratic field, $Q(\sqrt{d})$.

Galois was the first to use the term "adjoin" in a technical sense. The notion of adjoining the roots of an equation to the field of coefficients is central in his work.

One of the fundamental theorems of the subject proved by Galois is the Primitive Element Theorem. This says (in our terminology) that if E is the splitting field of a polynomial $f(x)$ over a field F, then $E = F(V)$ for some rational function V of the roots of $f(x)$. Galois used this result to determine the Galois group of the equation $f(x) = 0$. The Primitive Element Theorem was essential in all subsequent work in Galois theory until Artin bypassed it in the 1930s by reformulating Galois theory, for he felt that the theorem was not intrinsic to the subject.

4.2 Algebraic number theory

The central field-theoretic notion here, due independently to Dedekind and Kronecker, is that of an algebraic number field $Q(a)$, where a is an algebraic number. How did it arise? Mainly from three major number-theoretic problems: Fermat's Last Theorem (FLT), reciprocity laws, and representation of integers by binary quadratic forms. Although all three problems have to do with the domain of (ordinary) integers, in order to deal with them effectively it was found necessary to embed them in domains of what came to be known as algebraic integers. The following examples illustrate the ideas involved.

(a) To prove FLT for (say) $n = 3$, that is, to show that $x^3 + y^3 = z^3$ has no nonzero integer solutions, one factors the left side to obtain $(x + y)(x + yw)(x + yw^2) = z^3$, where w is a primitive cube root of unity, $w = (-1 + \sqrt{3}i)/2$. This is now an equation in the domain $D = \{a + bw : a, b \in Z\}$ of algebraic integers. This approach to FLT was essentially used by Euler (for $n = 3$) and later by Lamé and others.

(b) Gauss' quadratic reciprocity law appeared in his *Disquisitiones Arithmeticae* of 1801. It says that $x^2 \equiv p$ (mod q) is solvable if and only if $x^2 \equiv q$ (mod p) is solvable, unless $p \equiv q \equiv 3$ (mod 4), in which case $x^2 \equiv p$ (mod q) is solvable if and only if $x^2 \equiv q$ (mod p) is not. Here p and q are odd primes.

Gauss and others tried to extend this result to "higher" reciprocity laws. For example, for cubic reciprocity one asks about the relationship between the solvability of $x^3 \equiv p$ (mod q) and $x^3 \equiv q$ (mod p). These higher reciprocity-type problems are much more difficult to deal with than quadratic reciprocity. Gauss remarked that:

> The previously accepted laws of arithmetic are not sufficient for the foundations of a general theory [of higher reciprocity] ...Such a theory demands that the domain of arithmetic be endlessly enlarged.

Gauss' comments were no idle speculation. In fact, he himself began to implement the above "program" by formulating and proving a law of *biquadratic reciprocity*. To do that he extended the domain of arithmetic by introducing what came to be known as the *Gaussian integers* $G = \{a + bi : a, b \in Z\}$. He could not even *formulate* such a law without introducing G.

(c) The problem of representing integers by binary quadratic forms, namely determining the integers n for which $n = ax^2 + bxy + cy^2$ (for fixed $a, b, c \in Z$), goes back to Fermat. In particular, Fermat asked and answered the question: which integers n are sums of two squares, $n = x^2 + y^2$? In the *Disquisitiones* Gauss studied the *general* problem very thoroughly, developing a comprehensive and beautiful, but very difficult, theory. To gain deeper understanding of Gauss' theory of binary quadratic forms, Dedekind found that he too needed to extend the domain Z of integers. For example, even in the simple case of representing integers as sums of two squares, it is the equation $(x + yi)(x - yi) = z^2$ rather than $x^2 + y^2 = z^2$ that yields conceptual insight. See Chapter 3 for details.

4.2.1 Dedekind's Ideas

The fundamental question in extending the domain of ordinary arithmetic to "higher" domains is whether such domains "behave" like the integers, namely whether they are unique factorization domains (UFDs). It is this property that facilitates the solution of problems (a)–(c).

While the domains D and G introduced above *are* UFDs, most domains which arise in connection with the three number-theoretic problems we have described are not. For example, when we factor the left side of $x^p + y^p = z^p$ for $p \geq 23$ (p a prime), the resulting domains are *never* UFDs. To rescue unique factorization in such domains Dedekind introduced (in 1871) *ideals* and *prime ideals*, and showed that every ideal in these domains is a unique product of prime ideals. See Chapter 3.2.

But what *are* the domains with restored unique factorization? To answer that—one of the fundamental questions of his theory—Dedekind needed to introduce fields, in particular *algebraic number fields* $Q(a) = \{q_0 + q_1 a + q_2 a^2 + \cdots + q_n a^n : q_i \in Q\}$, where a is a root of a polynomial with integer coefficients. These were the natural habitats of his domains, just as the rationals are the natural habitat of the integers. The domains in question were then defined as "the integers of $Q(a)$," namely those elements of $Q(a)$ which are roots of *monic* polynomials with integer coefficients. Dedekind showed that they form a commutative ring with identity whose field of quotients is $Q(a)$.

Given Dedekind's predisposition for abstraction—a rather rare phenomenon in the 1870s, he placed his theory in a broader context by giving axiomatic definitions of rings, fields, and ideals. Here is his definition of a field:

> By a field we will mean every infinite system of real or complex numbers
> so closed in itself and perfect that addition, subtraction, multiplication, and
> division of any two of these numbers again yields a number of the system.

To Dedekind, then, fields were subsets of the complex numbers, which is, of course, all he needed for his theory of algebraic numbers. Still, an axiomatic definition in number theory/algebra, even in this restricted sense, is remarkable for that time. Also remarkable are Dedekind's use of infinite sets ("systems"), which predates Cantor's, and his "descriptive" rather than "constructive" definition of a mathematical object as a set of all elements of a certain kind satisfying a number of properties. See Chapter 8.2.

The field concept was a unifying mathematical notion for Dedekind. Before his definition of a field he says:

> In the following paragraphs I have attempted to introduce the reader into
> a higher domain, in which algebra and number theory interconnect in the
> most intimate manner.... I became convinced that studying the algebraic
> relationship of numbers is most conveniently based on a concept that is
> directly connected with the simplest arithmetic properties. I had originally
> used the term "rational domain," which I later changed to "field."

Hilbert remarked that Gauss, Dirichlet, and Jacobi had also expressed their amazement at the close connection between number theory and algebra, on the grounds that these subjects have common roots in (as Dedekind would put it) the theory of fields.

Dedekind produced several editions of his groundbreaking theory of ideal decomposition in algebraic number fields. In his mature 1894 version (4th edition of Dirichlet's *Zahlentheorie*) he included important concepts and results on fields—nowadays standard—such as:

(i) If S is any subset of the complex numbers containing the rationals, the intersection of all fields containing S is a field; it is called "rational with respect to S."

(ii) He defined field isomorphism, calling it "permutation of the field," as a mapping of a field E onto a field F which preserves all four operations of the field. He observed that if F is nonzero, the mapping is one-one. He also noted that the mapping is the identity on Q.

(iii) If E is a subfield of K, he defined the *degree* of K over E as the dimension of K considered as a vector space over E. He showed that if the degree is finite, then every element of K is algebraic over E.

4.2.2 Kronecker's Ideas

Kronecker's work was broader and much more difficult than Dedekind's. He developed his ideas over several decades, beginning in the 1850s, trying to frame a general theory which would subsume algebraic number theory and algebraic geometry as special cases. In his great 1882 work "Foundations of an arithmetic theory of algebraic numbers" he developed algebraic number theory using an entirely different approach from Dedekind's. One of his central concepts was also that of a field—he called it "domain of rationality," defined as follows:

> The domain of rationality $(R', R'', R''', \ldots)$ contains ... every one of those quantities which are rational functions of the quantities $R', R'', R''', \ldots$ with integer coefficients.

Leopold Kronecker (1823–1891)

Note how different Kronecker's "definition" of a field is from Dedekind's! It is a constructive description, rather than the kind of definition that would be acceptable to us today. But it was dictated by Kronecker's views on the nature of mathematics.

Kronecker rejected irrational numbers as bona fide entities since they involve the mathematical infinite. For example, the algebraic number field $Q(\sqrt{2})$ was defined by Kronecker as the quotient field of the polynomial ring $Q[x]$ relative to the ideal generated by $x^2 - 2$, though he would have put it in terms of congruences rather than quotient rings. These ideas contain the germ of what came to be known as *Kronecker's Theorem*, namely that every polynomial over a field has a root in some extension field.

It is interesting to compare the above definition of $Q(\sqrt{2})$ with Cauchy's definition in the 1840s of the complex numbers as polynomials over the reals modulo $x^2 + 1$ (and compare the latter with Gauss' integers modulo p). Cauchy's rationale was to give an "algebraic" definition of complex numbers which would avoid the use of $\sqrt{-1}$.

4.2.3 Dedekind vs Kronecker

Dedekind and Kronecker were great contemporary algebraists. Both published path-breaking works on algebraic number theory. But their approaches to the subject were very different. Both were guided in their works by their "philosophies" of mathematics, and these too were very different. Kronecker was perhaps the first preintuitionist, Dedekind likely the first preformalist (some have said he was the first logicist). Compare Kronecker's "God made the [positive] integers, all the rest is the work of man" with Dedekind's "[The natural] numbers are a free creation of the human mind." To Kronecker mathematics had to be constructive and finitary. Dedekind did not hesitate to use axiomatic notions and the infinite. While Kronecker made frequent pronouncements on these topics, Dedekind made few; his views became known mainly from his works—conceptual and abstract. Some examples:

(i) Since Kronecker's domains of rationality had to be generated by *finitely* many elements (the R', R'', R''', ...), his definition would not admit the totality of algebraic numbers as a field. Dedekind had no problem in considering the set of all complex numbers which are roots of polynomial equations with integer coefficients (viz. the set of all algebraic numbers) as a bona fide mathematical object.

(ii) On the other hand, Kronecker put no restriction on the *nature* of the entities R', R'', R''', ...—they could, for example, be indeterminates or roots of algebraic equations. So $Q(x)$ was a legitimate field to Kronecker. In fact, the adjunction of indeterminates to a field was a cornerstone of his approach to algebraic number theory. Dedekind, recall, defined his fields to be subsets of the complex numbers.

(iii) Since Kronecker did not accept π (say) as a legitimate number, he identified $Q(\pi)$ with $Q(x)$ (x an indeterminate), thus claiming that transcendental numbers are indeterminate. To Dedekind $Q(\pi)$ was a perfectly legitimate entity not requiring any assistance from $Q(x)$.

4.3 Algebraic geometry

The examples of fields we have come across so far have been mainly fields of *numbers*. Here we encounter principally fields of *functions*, in particular algebraic functions and rational functions. The ideas are due mainly to Kronecker and Dedekind–Weber.

4.3.1 Fields of Algebraic Functions

Algebraic geometry is the study of algebraic curves and their generalizations to higher dimensions, algebraic varieties. An *algebraic curve* is the set of roots of an algebraic function, that is, a function $y = f(x)$ defined implicitly by a polynomial equation $P(x, y) = 0$.

Several approaches were used in the study of algebraic curves, notably the analytic, the geometric-algebraic, and the algebraic-arithmetic. In the analytic approach, to which Riemann (in the 1850s) was the major contributor, the main objects of study were algebraic functions $f(w, z) = 0$ of a complex variable and their integrals, the so-called abelian integrals. It was in this connection that Riemann introduced the fundamental notion of a Riemann surface, on which algebraic functions become single-valued. Riemann's methods, however, were nonrigorous, relying heavily on the physically obvious but mathematically questionable Dirichlet Principle.

Dedekind and Weber, in their important 1882 paper "Theory of algebraic functions of a single variable," set for themselves the task of making Riemann's ideas rigorous. They put it thus:

> The purpose of the[se] investigations...is to justify the theory of algebraic functions of a single variable, which is one of the main achievements of Riemann's creative work, from a simple as well as rigorous and completely general viewpoint.

To accomplish this, they carried over to algebraic functions the ideas which Dedekind had earlier introduced for algebraic numbers. Specifically, just as an algebraic number field is a finite extension $Q(a)$ of the field Q of rational numbers, so Dedekind and Weber defined an *algebraic function field* as a finite extension $K = C(z)(w)$ of the field $C(z)$ of rational functions (in the indeterminate z). That is, w is a root of a polynomial $p(t) = a_0 + a_1t + a_2t^2 + \cdots + a_nt^n$, where $a_i \in C(z)$ (we can take $a_i \in C[z]$). Thus $w = f(z)$ is an algebraic function defined implicitly by the polynomial equation $P(z, w) = a_0 + a_1w + a_2w^2 + \cdots + a_nw^n = 0$. In fact, *all* the elements of $K = C(z)(w) = C(z, w)$ are algebraic functions.

Now let A be "the integers of K"; that is, A consists of the elements of $K = C(z)(w)$ which are roots of *monic* polynomials over $C[z]$ (cf. "the integers of $Q(a)$," p. 66). By analogy with the case of algebraic numbers, here too every nonzero ideal of A is a unique product of prime ideals. Incidentally, the meromorphic functions on a Riemann surface form a field of algebraic functions, with the entire functions as their "integers."

Dedekind and Weber were now ready to give a rigorous, *algebraic* definition of a Riemann surface S of the algebraic function field K: it is (in our terminology) the set of nontrivial discrete valuations on K. Many of Riemann's ideas on algebraic functions were here developed algebraically and rigorously.

Dedekind and Weber were at heart algebraists. They felt that algebraic function theory is intrinsically an algebraic subject, hence ought to be developed *algebraically*. As they put it: "In this way, a well-delimited and relatively comprehensive part of the theory of algebraic functions is treated solely by means belonging to its own domain."

Beyond their technical achievements in putting major parts of Riemann's algebraic function theory on solid ground, their conceptual breakthrough lay in pointing to the strong analogy between algebraic number fields and algebraic function fields, hence between algebraic number theory and algebraic geometry. This analogy proved most fruitful for both theories. Another noteworthy aspect of their work was its generality, in particular its applicability to arbitrary fields.

4.3.2 Fields of Rational Functions

As noted, algebraic geometry is the study of algebraic varieties. An *algebraic variety* is the set of points in $\mathbf{R}^n$ (or $\mathbf{C}^n$) satisfying a system of polynomial equations $f_i(x_1, x_2, \ldots, x_n) = 0, i = 1, 2, \ldots, k$. The ideal structure of the ring $\mathbf{R}[x_1, \ldots, x_n]$ (or $\mathbf{C}[x_1, \ldots, x_n]$) to which the polynomials $f_i(x_1, x_2, \ldots, x_n)$ belong is fundamental for the understanding of the algebraic variety, as is the "natural habitat" of that ring—its field of quotients $\mathbf{R}(x_1, \ldots, x_n)$ (or $\mathbf{C}(x_1, \ldots, x_n)$). These are the fields of (formally) *rational functions*. We have seen that such fields were also introduced by Kronecker in connection with his work in algebraic number theory (p. 67).

4.4 Congruences

Gauss introduced the congruence notation in the *Disquisitiones Arithmeticae* of 1801 and showed (among other things) that one can add, subtract, multiply, and divide congruences modulo a prime p, in effect that the integers modulo p form a field—a *finite field* of p elements. Inspired by Gauss' work on congruences, Galois introduced finite fields with p^n elements in an 1830 paper entitled "On the theory of numbers."

Galois' aim was to study the congruence $F(x) \equiv 0 \pmod{p}$ as a generalization of Gauss' quadratic congruences (cf. Gauss' quadratic reciprocity law, p. 65). Here $F(x)$ is a polynomial of degree n irreducible mod p, that is, irreducible over the field Z_p. Galois showed that $F(x)$ has neither repeated roots nor integral or rational roots. His conclusion was that:

> One should therefore regard the roots of this congruence as some kind of imaginary symbols ..., symbols whose employment in calculation will often prove as useful as that of the imaginary $\sqrt{-1}$ in ordinary analysis.

He continued:

> Let i [an arbitrary symbol, *not* the complex number i] denote one of the roots of the congruence $F(x) \equiv 0$, which can be supposed to have degree n. Consider the general expression
>
> $$a + a_1 i + a_2 i^2 + \cdots + a_{n-1} i^{n-1} (**),$$
>
> where $a, a_1, a_2, \ldots, a_{n-1}$ represent integers [mod p]. When these numbers are assigned all their possible values, expression $(**)$ runs through p^n values, which possess, as I shall demonstrate, the same properties as the natural numbers in the *theory of residues of powers*.

Galois did, indeed, show that the expressions $(**)$ form a field, now called a *Galois field*. He also showed that (in our terminology) the multiplicative group of that field is cyclic. In an 1893 paper entitled "A doubly-infinite system of simple groups," E.H. Moore characterized the finite fields [15].

4.5 Symbolical algebra

In the third and fourth decades of the nineteenth century British mathematicians, notably Peacock, Gregory, and De Morgan, created what came to be known as symbolical algebra. Their aim was to set algebra—to them this meant the laws of operation with numbers, negative numbers especially—on an equal footing with geometry by providing it with logical justification. They did this by distinguishing between *arithmetical algebra*—laws of operation with *positive* numbers, and *symbolical algebra*—a subject newly created by Peacock which dealt with laws of operation with numbers in general.

Although the laws were carried over verbatim from those of arithmetical algebra, in accordance with the so-called *Principle of Permanence of Equivalent Forms*, the point of view was remarkably modern. Witness Peacock's definition of symbolical algebra, given in his *Treatise of Algebra*, as

> The science which treats of the combinations of arbitrary signs and symbols by means of defined though arbitrary laws.

Quite a statement for the early nineteenth century! Such sentiments were about a century ahead of their time. And of course one *did* have to wait about a century to have what Peacock had preached put fully into practice. Nevertheless, the creation of symbolical algebra was a significant development, even if not *directly* related to fields, signalling (according to some) the birth of abstract algebra. Moreover, although Peacock did not specify the nature of the "arbitrary laws," they turned later in the century into axioms for rings and fields. See Chapter 1.8 for further details.

4.6 The abstract definition of a field

The developments we have been describing thus far lasted close to a century. They gave rise to important "concrete" theories—Galois theory, algebraic number theory, algebraic geometry—in which the (at times implicit) field concept played a central role.

At the end of the nineteenth century abstraction and axiomatics were "in the air." For example, Pasch (1882) gave axioms for projective geometry, stressing for the first time the importance of undefined notions, Cantor (1883) defined the real numbers as equivalence classes of Cauchy sequences of rationals, and Peano (1889) gave his axioms for the natural numbers. In algebra, von Dyck (1882) gave an abstract definition of a group which encompassed both finite and infinite groups (about thirty years earlier Cayley had defined a *finite* group), and Peano (1888) gave a definition of a finite-dimensional vector space, though this was largely ignored by his contemporaries. The time was propitious for the abstract field concept to emerge. Emerge it did in 1893 in the hands of Weber (of Dedekind–Weber fame).

Weber's definition of a field appeared in his 1893 paper "General foundations of Galois' theory of equations" [23], in which he aimed to give an abstract formulation of Galois theory:

> In the following an attempt is made to present the Galois theory of algebraic equations in a way which will include equally well all cases in which this theory might be used. Thus we present it here as a direct consequence of the group concept illuminated by the field concept, as a formal structure completely without reference to any numerical interpretation of the elements used.

Heinrich Weber (1842–1913)

Weber's presentation of Galois theory is indeed very close to the way the subject is taught today. His definition of a field, preceded by that of a group, is as follows:

> A group becomes a field if two types of composition are possible in it, the first of which may be called *addition*, the second *multiplication*. The general determination must be somewhat restricted, however.
>
> 1. We assume that both types of composition are commutative.
> 2. Addition shall generally satisfy the conditions which define a group.
> 3. Multiplication is such that
>
> $$a(-b) = -(ab)$$
> $$a(b + c) = ab + ac$$
> $$ab = ac \quad \text{implies } b = c, \text{ unless } a = 0$$
> Given b and c, $ab = c$ determines a, unless $b = 0$.

Although the associative law under multiplication is missing, and the axioms are not independent, they are of course very much in the modern spirit. As examples of his newly-defined concept Weber included the number fields and function fields of algebraic number theory and algebraic geometry, respectively, but also Galois' finite fields and Kronecker's "congruence fields" $K[x]/(p(x))$, K a field, $p(x)$ a polynomial irreducible over K.

Weber proved (often reproved, after Dedekind) various theorems about fields which later became useful in Artin's formulation of Galois theory, and which are today recognized as basic results of the theory. Among them are the following:

(i) Every finite algebraic extension of a field is simple (that is, is generated by a single element).

(ii) Every polynomial over a field has a splitting field.

(iii) If $F \subseteq F(a) \subseteq F(b)$, then $(F(a) : F)$ divides $(F(b) : F)$, where for fields K and E with $E \subseteq K$, $(K : E)$ denotes the dimension of K as a vector space over E.

It should be emphasized that it was not Weber's aim to study fields as such, but rather to develop enough of field theory to give an abstract formulation of Galois theory. In this he succeeded admirably. His paper, and somewhat later his two-volume *Textbook on Algebra*, exerted considerable influence on the development of abstract algebra.

4.7 Hensel's *p*-adic numbers

In an 1899 article entitled "New foundations of the theory of algebraic numbers," Hensel began a life-long study of p-adic numbers. Inspired by the work of Dedekind–Weber we have described above, Hensel took as his point of departure the analogy between function fields and number fields (p. 55). Just as power series are useful for a study of the former, Hensel introduced p-adic numbers to aid in the study of the latter:

> The analogy between the results of the theory of algebraic functions of one variable and those of the theory of algebraic numbers suggested to me many years ago the idea of replacing the decomposition of algebraic numbers, with the help of ideal prime factors, by a more convenient procedure that fully corresponds to the expansion of an algebraic function in power series in the neighborhood of an arbitrary point.

Indeed, in the neighborhood of a given point α every algebraic function of a complex variable can be represented as an infinite series of integral and rational powers of $z - \alpha$, as Weierstrass had shown. The elements of Hensel's *field of p-adic numbers* are formal power series $\sum a_k p^k$, where $a_k \in Z_p$ and $k \in Z$, with finitely many negative exponents. (Formal power series were introduced by Veronese in a geometric context in 1891.) And just as every element of an algebraic function field can be identified with the set of its expansions at all points α of the Riemann surface on which it is defined, so every element of an algebraic number field is identified with the set of its representations in the field of p-adic numbers $\sum a_k p^k$ for every prime p.

In a book of 1907 Hensel introduced topological notions in his p-adic fields and applied the resulting p-adic analysis in algebraic number theory. The p-adic numbers proved extremely useful also in algebraic geometry. And they were influential in motivating the abstract study of rings and fields.

4.8 Steinitz

The last major event in the evolution of field theory that we want to describe is Steinitz's great work of 1910 [20]. But first some background.

Algebra in the nineteenth century was by our standards concrete. It was connected in one way or another with the real or complex numbers. For example, some of the great contributors to nineteenth-century algebra, mathematicians whose ideas shaped the algebra of the twentieth century, were Gauss, Galois, Jordan, Kronecker, Dedekind, and Hilbert. Their algebraic work dealt with quadratic forms, cyclotomy, permutation groups, ideals in rings of algebraic number fields and algebraic function fields, and invariant theory. All of these subjects were related in one way or another to the real or complex numbers.

At the turn of the twentieth century the axiomatic method began to take hold as an important mathematical tool. Hilbert's *Foundations of Geometry* of 1899 was very influential in this respect. Noteworthy also was the American school of axiomatic analysis, as exemplified in the works of Dickson, Huntington, E. H. Moore, and Veblen. In the first decade of the twentieth century these mathematicians began to examine various axiom systems for groups, fields, associative algebras, projective geometry, and the algebra of logic. Their principal aim was to study the independence, consistency, and completeness of the axioms defining any one of these systems (see [25]). Also relevant were Hilbert's axiomatic characterization in 1900 of the field of real numbers and Huntington's like characterization in 1905 of the field of complex numbers. See [2], [4] for details.

Steinitz's groundbreaking 150-page work "Algebraic theory of fields" of 1910 initiated the abstract study of fields as an independent subject [20]. While Weber *defined* fields abstractly, Steinitz *studied* them abstractly.

Steinitz's immediate source of inspiration was Hensel's p-adic numbers:

> I was led into this general research especially by Hensel's *Theory of Algebraic Numbers*, whose starting point is the field of p-adic numbers, a field which counts neither as a field of functions nor as a field of numbers in the usual sense of the word.

More generally, Steinitz's work arose out of a desire to delineate the abstract notions common to the various contemporary theories of fields: fields in algebraic number theory, in algebraic geometry, and in Galois theory, p-adic fields, and finite fields. His goal was a comprehensive study of *all* fields, starting from the field axioms:

> The aim of the present work is to advance an overview of all the possible types of fields and to establish the basic elements of their interrelations.

Ernst Steinitz (1871–1928)

Quite a task! Steinitz's plan was to start from the simplest fields and to build up all fields from these. The basic concept which he identified to study the former is the *characteristic* of the field. Here are several of his fundamental results, nowadays staples of field theory:

(i) Classification of fields into those of characteristic zero and those of characteristic p. The *prime fields*—the "simplest" fields—are Q and Z_p; one or the other is a subfield of every field.
(ii) Development of a theory of *transcendental extensions*, which became indispensable in algebraic geometry.
(iii) Recognition that it is precisely the *finite, normal, separable extensions* to which Galois theory applies.
(iv) Proof of the existence and uniqueness (up to isomorphism) of the *algebraic closure* of any field.

A description of all fields followed:

> Starting with an arbitrary prime field, by taking an arbitrary, purely transcendental extension followed by an arbitrary algebraic extension, we have a method of arriving at any field.

The notions of *transcendency base* and *degree of transcendence* of an extension field, both of which Steinitz introduced, played a crucial role here. Also important was the axiom of choice, whose use he acknowledged:

> Many mathematicians continue to reject the axiom of choice. The growing realization that there are questions in mathematics that cannot be decided

without this principle is likely to result in the gradual disappearance of the resistance to it.

Steinitz's work was very influential in the development of abstract algebra in the 1920s and 1930s, as the following testimonials show:

> Steinitz's paper was the basis for all [algebraic] investigations in the school of Emmy Noether (van der Waerden [22]).

> [Steinitz's work] … is not only a landmark in the development of algebra, but also … an excellent, in fact indispensable, introduction to a serious study of the new [modern] algebra (Baer & Hasse [20]).

> Steinitz's work marks a methodological turning point in algebra, leading to … 'modern' algebra (Purkert & Wussing [17]).

> [Steinitz's work] can be considered as having given birth to the actual concept of Algebra (Bourbaki [3]).

4.9 A glance ahead

Below we list several major developments in field theory and related areas in the decades following Steinitz's fundamental work.

(a) *Valuation theory.* In 1913 Kürschak abstracted Hensel's ideas on p-adic fields by introducing the notion of a *valuation field.* He proved the existence of the *completion* of a field with respect to a valuation. In 1918 Ostrowski determined all valuations of the field Q of rational numbers. Valuation theory, which "forms a solid link between number theory, algebra, and analysis," according to Jacobson [10], played fundamental roles in both algebraic number theory and algebraic geometry. See [3], [6], [10], [22].

(b) *Formally real fields.* In 1927 Artin and Schreier defined the notion of a *formally real field*, namely a field in which -1 is not a sum of squares. According to Bourbaki,

> One of [the] remarkable results [of the Artin–Schreier theory] is no doubt the discovery that the existence of an order relation on a field is linked to purely algebraic properties of the field.

Specifically, a field can be ordered if and only if it is formally real. The theory of formally real fields enabled Artin in the same year to solve *Hilbert's* 17th *Problem* on the resolution of positive definite rational functions into sums of squares.

(c) *Class field theory.* This is the study of finite extensions of an algebraic number field having an abelian Galois group. It is a beautiful synthesis of algebraic, number-theoretic, and analytic ideas, in which *Artin's Reciprocity Law* has a central place. Major strides were already made by Hilbert in his "Zahlbericht" ("Report on Number Theory") of 1897. More modern aspects of the theory were developed by Artin, Chevalley, Hasse, Tagaki, and others. See [8].

(d) *Galois theory*. Artin set out his now famous abstract formulation of Galois theory in lectures given in 1926 (but published only in 1938). In a 1950 talk he said:

> Since my mathematical youth I have been under the spell of the classical theory of Galois. This charm has forced me to return to it again and again, and try to find new ways to prove its fundamental theorems.

Extensions of the classical theory were given in various directions. For example, in 1927 Krull developed a Galois theory of *infinite field extensions*, establishing a one-one correspondence between subfields and "closed" subgroups, and thereby introducing topological notions into the theory. There is also a Galois theory for *inseparable field extensions*, in which the notion of derivation of a field plays a central role, and a Galois theory for *division rings*, developed independently by H. Cartan and Jacobson in the 1940s. See [10], [24].

(e) *Finite fields*. Finite field theory is a thriving subject of investigation in its own right, but it also has important uses in number theory, coding theory, geometry, and combinatorics. See [9], [14].

References

1. I. G. Bashmakova and E. I. Slavutin, Algebra and algebraic number theory, in *Mathematics of the 19th Century*, ed. by A. N. Kolmogorov and A. P. Yushkevich, Birkhäuser, 1992, pp. 35–135.
2. G. Birkhoff, Current trends in algebra, *American Math. Monthly* 1973, **80**: 760–782, and corrections in 1974, **81**: 746.
3. N. Bourbaki, *Elements of the History of Mathematics*, Springer-Verlag, 1984.
4. L. Corry, *Modern Algebra and the Rise of Mathematical Structures*, Birkhäuser, 1996.
5. H. M. Edwards, *Fermat's Last Theorem: A Genetic Introduction to Algebraic Number Theory*, Springer-Verlag, 1977.
6. D. Eisenbud, *Commutative Algebra with a View Toward Algebraic Geometry*, Springer-Verlag, 1995.
7. E. Galois, Sur la théorie des nombres. English translation in *Introductory Modern Algebra: A Historical Approach*, by S. Stahl, Wiley, 1997, pp. 277–284.
8. H. Hasse, History of class field theory, in *Algebraic Number Theory, Proceedings of an Instructional Conference*, ed. by J. Cassels & A. Fröhlich, Thompson Book Co., 1967, pp. 266–279.
9. K. Ireland and M. Rosen, *A Classical Introduction to Modern Number Theory*, 2nd ed., Springer-Verlag, 1982.
10. N. Jacobson, *Basic Algebra I, II*, W. H. Freeman, 1974 & 1980.
11. B. M. Kiernan, The development of Galois theory from Lagrange to Artin, *Arch. Hist. Ex. Sc.* 1971/72, **8**: 40–154.
12. I. Kleiner, The roots of commutative algebra in algebraic number theory, *Math. Mag.* 1995, **68**: 3–15.
13. D. Laugwitz, *Bernhard Riemann, 1826–1866*, Birkhäuser, 1999. (Translated from the German by A. Shenitzer.)
14. R. Lidl and H. Niederreiter, *Introduction to Finite Fields and their Applications*, Cambridge University Press, 1986.

15. E. H. Moore, A doubly-infinite system of simple groups, *New York Math. Soc. Bull.* 1893, **3**: 73–78.
16. W. Purkert, Zur Genesis des abstrakten Körperbegriffs I, II, *NTM* 1971, **8**: 23–37 and 1973, **10**: 8–20. (Unpublished English translation by A. Shenitzer.)
17. W. Purkert and H. Wussing, Abstract algebra, in *Companion Encyclopedia of the History and Philosophy of the Mathematical Sciences*, ed. by I. Grattan-Guinness, Routledge, 1994, vol. 1, pp. 741–760.
18. H. M. Pycior, George Peacock and the British origins of symbolical algebra, *Hist. Math.* 1981, **8**: 23–45.
19. J. H. Silverman and J. Tate, *Rational Points on Elliptic Curves*, Springer-Verlag, 1992.
20. E. Steinitz, *Algebraische Theorie der Körper*, 2nd ed., Chelsea, 1950.
21. J.-P. Tignol, *Galois' Theory of Algebraic Equations*, Wiley, 1988.
22. B. L. van der Waerden, Die Algebra seit Galois, *Jahresbericht d. DMV* 1966, **68**: 155–165.
23. H. Weber, Die allgemeinen Grundlagen der Galois'schen Gleichungstheorie, *Math. Ann.* 1893, **43**: 521–549.
24. D. Winter, *The Structure of Fields*, Springer-Verlag, 1974.
25. M. Scanlan, Who were the American postulate theorists?, *Jour. of Symbolic Logic* 1991, **56**: 981–1002.

5

History of Linear Algebra

Linear algebra is a very useful subject, and its basic concepts arose and were used in different areas of mathematics and its applications. It is therefore not surprising that the subject had its roots in such diverse fields as number theory (both elementary and algebraic), geometry, abstract algebra (groups, rings, fields, Galois theory), analysis (differential equations, integral equations, and functional analysis), and physics. Among the elementary concepts of linear algebra are linear equations, matrices, determinants, linear transformations, linear independence, dimension, bilinear forms, quadratic forms, and vector spaces. Since these concepts are closely interconnected, several usually appear in a given context (e.g., linear equations and matrices) and it is often impossible to disengage them.

By 1880, many of the basic results of linear algebra had been established, but they were not part of a general theory. In particular, the fundamental notion of vector space, within which such a theory would be framed, was absent. This was introduced only in 1888 by Peano. Even then it was largely ignored (as was the earlier pioneering work of Grassmann), and it took off as the essential element of a fully-fledged theory in the early decades of the twentieth century. So the historical development of the subject is the reverse of its logical order.

We will describe the elementary aspects of the evolution of linear algebra under the following headings: linear equations; determinants; matrices and linear transformations; linear independence, basis and dimension; and vector spaces. Along the way, we will comment on some of the other concepts mentioned above.

5.1 Linear equations

About 4000 years ago the Babylonians knew how to solve a system of two linear equations in two unknowns (a 2×2 system). In their famous *Nine Chapters of the Mathematical Art* (c. 200 BC) the Chinese solved 3×3 systems by working solely with their (numerical) coefficients. These were prototypes of matrix methods, not unlike the "elimination methods" introduced by Gauss and others some 2000 years later. See [20].

The modern study of systems of linear equations can be said to have originated with Leibniz, who in 1693 invented the notion of a determinant for this purpose. But his investigations remained unknown at the time. In his *Introduction to the Analysis of Algebraic Curves* of 1750, Cramer published the rule named after him for the solution of an $n \times n$ system, but he provided no proofs. He was led to study systems of linear equations while attempting to solve a geometric problem, determining an algebraic curve of degree n passing through $(1/2)n^2 + (3/2)n$ fixed points. See [1], [20].

Gottfried Wilhelm Leibniz (1646–1716)

Euler was perhaps the first to observe that a system of n equations in n unknowns does not necessarily have a unique solution, noting that to obtain uniqueness it is necessary to add conditions. He had in mind the idea of dependence of one equation on the others, although he did not give precise conditions. In the eighteenth century the study of linear equations was usually subsumed under that of determinants, so no consideration was given to systems in which the number of equations differed from the number of unknowns. See [8], [9].

In connection with his invention of the method of least squares (published in a paper in 1811 dealing with the determination of the orbit of an asteroid), Gauss introduced a systematic procedure, now called *Gaussian elimination*, for the solution of systems of linear equations, though he did not use the matrix notation. He dealt with the cases in which the number of equations and unknowns may differ [20]. The theoretical properties of systems of linear equations, including the issue of their consistency, were treated in the second half of the nineteenth century, and were at least partly motivated by questions of the reduction of quadratic and bilinear forms to "simple" (canonical) ones. See [16], [18].

5.2 Determinants

Although one speaks nowadays of the determinant of a matrix, the two concepts had different origins. In particular, determinants appeared before matrices, and the early stages in their history were closely tied to linear equations. Subsequent problems that gave rise to new uses of determinants included *elimination theory* (finding conditions under which two polynomials have a common root), transformation of coordinates to simplify algebraic expressions (e.g., quadratic forms), change of variables in multiple integrals, solution of systems of differential equations, and celestial mechanics. See [24].

As we have noted in the previous section on linear equations, Leibniz invented determinants. He "knew in substance the[ir] modern combinatorial definition" [21], and he used them in solving linear equations and in elimination theory. He wrote many papers on determinants, but they remained unpublished till recently. See [21], [22].

The first publication to contain some elementary information on determinants was Maclaurin's *Treatise of Algebra*, in which they were used to solve 2×2 and 3×3 systems. This was soon followed by Cramer's significant use of determinants (cf. the previous section). See [1], [20], [21].

An exposition of the theory of determinants independent of their relation to the solvability of linear equations was first given by Vandermonde in his "Memoir on elimination theory" of 1772. (The word "determinant" was used for the first time by Gauss, in 1801, to stand for the discriminant of a quadratic form, where the discriminant of the form $ax^2 + bxy + cy^2$ is $b^2 - 4ac$.) Laplace extended some of Vandermonde's work in his *Researches on the Integral Calculus and the System of the World* (1772), showing how to expand $n \times n$ determinants by cofactors. See [24].

The first to give a systematic treatment of determinants was Cauchy in an 1815 paper entitled "On functions which can assume but two equal values of opposite sign by means of transformations carried out on their variables." He can be said to be the founder of the theory of determinants as we know it today. Many of the results on determinants found in a first textbook on linear algebra are due to him. For example, he proved the important product rule $\det(AB) = (\det A)(\det B)$. His work provided mathematicians with a powerful algebraic apparatus for dealing with n-dimensional algebra, geometry, and analysis. For instance, in 1843 Cayley developed the analytic geometry of n dimensions using determinants as a basic tool, and in the 1870s Dedekind used them to prove the important result that sums and products of algebraic integers are algebraic integers. See [18], [21], [22], [24].

Weierstrass and Kronecker introduced a definition of the determinant in terms of axioms, probably in the 1860s. (Rigorous thinking was characteristic of both mathematicians.) For example, Weierstrass defined the determinant as a normed, linear, homogeneous function. Their work became known in 1903, when Weierstrass' *On Determinant Theory* and Kronecker's *Lectures on Determinant Theory* were published posthumously. Determinant theory was a vigorous and independent subject of research in the nineteenth century, with over 2000 published papers. But it became largely unfashionable for much of the twentieth century, when determinants were no longer needed to prove the main results of linear algebra. See [21], [22], [24], [25].

5.3 Matrices and linear transformations

Matrices are "natural" mathematical objects: they appear in connection with linear equations, linear transformations, and also in conjunction with bilinear and quadratic forms, which were important in geometry, analysis, number theory, and physics.

Matrices as rectangular arrays of numbers appeared around 200 BC in Chinese mathematics, but there they were merely abbreviations for systems of linear equations. Matrices become important only when they are operated on—added, subtracted, and especially multiplied; more important, when it is shown what use they are to be put to.

Matrices were introduced implicitly as abbreviations of linear transformations by Gauss in his *Disquisitiones* mentioned earlier, but now in a significant way. Gauss undertook a deep study of the arithmetic theory of binary quadratic forms, $f(x, y) = ax^2 + bxy + cy^2$. He called two forms $f(x, y)$ and $F(X, Y) = AX^2 + BXY + CY^2$ "equivalent" if they yield the same set of integers, as x, y, X, and Y range over all the integers (a, b, c and A, B, C are integers). He showed that this is the same as saying that there exists a linear transformation T of the coordinates (x, y) to (X, Y) with determinant $= 1$ that transforms $f(x, y)$ into $F(X, Y)$. The linear transformations were represented as rectangular arrays of numbers—matrices, although Gauss did not use matrix terminology. He also defined implicitly the product of matrices (for the 2×2 and 3×3 cases only); he had in mind the composition of the corresponding linear transformations. See [1], [7], [16].

Linear transformations of coordinates, $y_j = \sum_{k=1}^{n} a_{jk} x_k (1 \leq j \leq m)$, appear prominently in the analytic geometry of the seventeenth and eighteenth centuries (mainly for $m = n \leq 3$). This led naturally to computations done on rectangular arrays of numbers (a_{jk}). Linear transformations also show up in projective geometry, founded in the seventeenth century and described analytically in the early nineteenth. See [2], [9].

In attempts to extend Gauss' work on quadratic forms, Eisenstein and Hermite tried to construct a general arithmetic theory of forms $f(x_1, x_2, \ldots, x_n)$ of any degree in any number of variables. In this connection they too introduced linear transformations, denoted them by single letters—an important idea—and studied them as independent entities, defining their addition and multiplication (composition). See [16].

Cayley formally introduced $m \times n$ matrices in two papers in 1850 and 1858 (the *term* "matrix" was coined by Sylvester in 1850). He noted that they "comport themselves as single entities" and recognized their usefulness in simplifying systems of linear equations and composition of linear transformations. He defined the sum and product of matrices for suitable pairs of rectangular matrices, and the product of a matrix by a scalar, a real or complex number. He also introduced the identity matrix and the inverse of a square matrix, and showed how the latter can be used in solving $n \times n$ linear systems under certain conditions.

In his 1858 paper "A memoir on the theory of matrices" Cayley proved the important Cayley–Hamilton theorem that a square matrix satisfies its characteristic polynomial. The proof consisted of computations with 2×2 matrices, and the observation that he had verified the result for 3×3 matrices. He noted that the result applies more widely. But he added: "I have not thought it necessary to undertake the labour of a formal proof of the theorem in the general case of a matrix of any degree." Hamilton proved the theorem independently (for $n = 4$, but without using the matrix notation) in his work on quaternions. Cayley used matrices in another paper to solve a significant problem, the so-called Cayley–Hermite problem, which asks for the determination of all linear transformations leaving a quadratic form in n variables invariant. See [16], [18].

Cayley advanced considerably the important idea of viewing matrices as constituting a symbolic algebra. In particular, his use of a single letter to represent a matrix was a significant step in the evolution of matrix algebra. But his papers of the 1850s were little noticed outside England until the 1880s. See [3], [4], [12], [16], [20], and Chapter 8.1.3.

During the intervening years (roughly the 1820s–1870s) deep work on matrices (in one guise or another) was done on the continent, by Cauchy, Jacobi, Jordan, Weierstrass, and others. They created what may be called the *spectral theory* of matrices: their classification into types such as symmetric, orthogonal, and unitary; results on the nature of the eigenvalues of the various types of matrices; and, above all, the theory of canonical forms for matrices—the determination, among all matrices of a certain type, of those that are canonical in some sense. An important example is the *Jordan canonical form*, introduced by Weierstrass (and independently by Jordan), who showed that two matrices are similar if and only if they have the same Jordan canonical form.

Spectral theory originated in the eighteenth century in the study of physical problems. This led to the investigation of differential equations and eigenvalue problems. In the nineteenth century these ideas gave rise to a purely mathematical theory. Hawkins gives an excellent account of this development in several articles [15], [16], [17], [18]; see also [11].

In a seminal paper in 1878 titled "On linear substitutions and bilinear forms" Frobenius developed substantial elements of the theory of matrices in the language of bilinear forms. (The theory of bilinear and quadratic forms was created by Weierstrass and Kronecker.) The forms, he said, can be viewed "as a system of n^2 quantities which are ordered in n rows and n columns." He was inspired by his teacher Weierstrass, and his paper is in the Weierstrass tradition of stressing a rigorous approach and seeking the fundamental ideas underlying the theories. For example, he made a thorough study of the general problem of canonical forms for bilinear forms, attributing special cases to Kronecker and Weierstrass. "Frobenius' paper ... represents an important landmark in the history of the theory of matrices, for it brought together for the first time the work on spectral theory of Cauchy, Jacobi, Weierstrass and Kronecker with the symbolical tradition of Eisenstein, Hermite and Cayley" [18]. See also [15], [16].

Frobenius applied his theory of matrices to group representations and to quaternions, showing for the latter that the only n-tuples of real numbers which are division algebras are the real numbers, the complex numbers, and the quaternions, a result proved independently by C. S. Peirce. (Cayley, in his 1858 paper, also related matrices to quaternions by showing that the quaternions are isomorphic to a subalgebra of the algebra of 2×2 matrices over the complex numbers.) The relationship

Georg Ferdinand Frobenius (1849–1917)

between (associative) algebras and matrices was to be of fundamental importance for subsequent developments of noncommutative ring theory. See Chapter 3.1.

5.4 Linear independence, basis, and dimension

The notions of linear independence, basis, and dimension appear, not necessarily with formal definitions, in various contexts, among them algebraic number theory, fields and Galois theory, hypercomplex number systems (algebras), differential equations, and analytic geometry.

In algebraic number theory the objects of study are algebraic number fields $Q(\alpha)$, where Q denotes the rationals and α is an algebraic number. If the minimal polynomial of α has degree n, then every element of $Q(\alpha)$ can be expressed uniquely in the form $a_0 + a_1\alpha + a_2\alpha^2 + \cdots + a_{n-1}\alpha^{n-1}$, where $a_i \in Q$. Thus $1, \alpha, \alpha^2, \ldots \alpha^{n-1}$ form a basis of $Q(\alpha)$, considered as a vector space over Q. This is precisely the line of thought pursued by Dedekind in his Supplement X of 1871 to Dirichlet's book on number theory (and with more clarity and detail in Supplements to subsequent editions of Dirichlet's book), although there is no formal definition of a vector space. See Chapter 3.2 and especially Chapter 8.2.4.

In connection with his work in algebraic number theory Dedekind introduced the notion of a field. He defined it as a subset of the complex numbers satisfying certain axioms. He included important concepts and results on fields, some related to ideas of linear algebra. For example, if E is a subfield of K, he defined the "degree" of

K over E as the dimension of K considered as a vector space over E, and showed that if the degree is finite, then every element of K is algebraic over E. The notions of linear independence, basis, and dimension appear here in a transparent way; the notion of vector space also appears, but only implicitly. See Chapter 4.2 and Chapter 8.2.

In 1893 Weber gave an axiomatic definition of finite groups and fields, with the objective of giving an abstract formulation of Galois theory. Among the results on fields is the following: If F is a subfield of E, which in turn is a subfield of K, then $(K : F) = (K : E)(E : F)$, where for any subfield S of T, $(T : S)$ denotes the dimension of T as a vector space over S (we assume that all dimensions in question are finite). See Chapter 4.6.

Many of the ideas of Dedekind and Weber on field extensions were brought to perfection in Steinitz's groundbreaking paper of 1910, "Algebraic theory of fields," in which he presented an abstract development of field theory. In the 1920s Artin "linearized" Galois theory—a most important idea. Contemporary treatment of the subject usually follows his. See [9] and Chapter 4.8.

An algebra (it is both a ring and a vector space over a field) was called in the nineteenth century a hypercomplex number system. The first such example was Hamilton's quaternions. This inspired generalizations to higher dimensions. For example, Cayley and Graves independently introduced octonions (in 1844), elements of the form $a_1e_1 + a_2e_2 + \cdots + a_8e_8$, where a_i are real numbers and e_i are "basis" elements subject to laws of multiplication. In 1854, in a paper in which he defined finite groups, Cayley introduced the group algebra of such a group—a linear combination of the group elements with real or complex coefficients. He called it a system of "complex quantities" and observed that it is analogous in many ways to the quaternions. See Chapter 3.1.

In a groundbreaking paper in 1870 entitled "Linear associative algebra," B. Peirce gave a definition of a finite-dimensional associative algebra as the totality of formal expressions of the form $\sum_{i=1}^{n} a_i e_i$, where the a_i are complex numbers and the e_i are "basis" elements, subject to associative and distributive laws. See Chapter 3.1.

Euler began to lay the framework for the solution of linear homogeneous differential equations. He observed that the general solution of such an equation with constant coefficients could be expressed as a linear combination of linearly independent particular solutions. Later in the eighteenth century Lagrange extended his result to equations with nonconstant coefficients. Demidov [6] discusses the analogy between linear algebraic equations and linear differential equations, focusing on the early nineteenth century.

The interaction of algebra and geometry is fundamental in linear algebra (for example, n-dimensional Euclidean geometry can be viewed as an n-dimensional vector space over the reals together with a symmetric bilinear form $B(x, y) = \sum a_{ij}x_i y_j$ that serves to define the length of a vector and the angle between two vectors). It began with the introduction of analytic geometry by Descartes and Fermat in the early seventeenth century, was extended by Euler in the eighteenth to 3 dimensions, and was put in modern form (the way we see it today) by Monge in the early nineteenth. The notions of linear combination, coordinate system, and basis were fundamental,

Leonhard Euler (1707–1783)

as were other basic notions of linear algebra such as matrix, determinant, and linear transformation. See [2], [8], [9], [13].

Further details on linear independence, basis, and dimension appear in the following section.

5.5 Vector spaces

As we mentioned, by 1880 many of the fundamental results of linear algebra had been established, but they were not considered as parts of a general theory. In particular, the fundamental notion of vector space, within which such a theory would be framed, was absent. It was introduced by Peano in 1888.

The earliest notion of *vector* comes from physics, where it means a quantity having both magnitude and direction (e.g., velocity or force). This idea was well established by the end of the seventeenth century, as was that of the parallelogram of vectors, a parallelogram determined by two vectors as its adjacent sides. In this setting the addition of vectors and their multiplication by scalars had clear physical meanings. See [5].

The *mathematical* notion of vector originated in the geometric representation of complex numbers, introduced independently by several authors in the late eighteenth and early nineteenth centuries, starting with Wessel in 1797 and culminating with Gauss in 1831. The representation of the complex numbers in these works was geometric, as points or as directed line segments in the plane. In 1835 Hamilton defined the complex numbers algebraically as ordered pairs of reals, with the usual operations of addition and multiplication, as well as multiplication by real numbers (the term "scalar" originated in his work on quaternions). He noted that these pairs satisfy the closure laws and the commutative and distributive laws, have a zero element, and have additive and multiplicative inverses. (The associative laws were mentioned in his 1843 work on quaternions.) See [5], [14].

An important development was the extension of vector ideas to three-dimensional space. Hamilton constructed an algebra of vectors within his system of quaternions. (Josiah Willard Gibbs and Oliver Heaviside introduced a competing system—their vector analysis—in the 1880s.) These were represented in the form $ai + bj + ck$, where a, b, c are real numbers and i, j, k the quaternion units—a clear precursor of a basis for three-dimensional Euclidean space. It was here that he introduced the term "vector" for these objects. See [5], [14].

A crucial development in vector-space theory was the further extension of the notions on three-dimensional space to spaces of higher dimension, advanced independently in the early 1840s by Cayley, Hamilton, and Grassmann. Hamilton called the extension of 3-space to four dimensions a "leap of the imagination." He had in mind, of course, his quaternions, a four-dimensional vector space (also a division algebra). He introduced them in dramatic fashion in 1843, and spent the next twenty years in their exploration and applications. Cayley's ideas on dimensionality appeared in his 1843 paper "Chapters of analytic geometry of n-dimensions." See [2], [5], [14], [19], and Chapter 8.5.

The pioneering ideas were expounded by Grassmann in his *Doctrine of Linear Extension* (1844). This was a brilliant work whose aim was to construct a coordinate-free algebra of n-dimensional space. It contained many of the basic ideas of linear algebra, including the notion of an n-dimensional vector space, subspace, spanning set, independence, basis, dimension, and linear transformation.

The definition of vector space was given as the set of linear combinations $\sum a_i e_i (i = 1, 2, \ldots, n)$, where a_i are real numbers and e_i "units," assumed to be linearly independent. Addition, subtraction, and multiplication by real numbers of such sums were defined in the usual manner, followed by a list of "fundamental properties." Among these are the commutative and associative laws of addition, the subtraction laws $a + b - b = a$ and $a - b + b = a$, and several laws dealing with multiplication by scalars. From these, Grassmann claimed, all the usual laws of addition, subtraction, and multiplication by scalars follow. He proved various results about vector spaces, including the fundamental relation $\dim V + \dim W = \dim(V + W) + \dim(V \cap W)$ for subspaces V and W of a vector space. See [5], [9], [10].

Grassmann's work was difficult to understand, containing many new ideas couched in philosophical language. It was thus ignored by the mathematical community. An 1862 edition was better received. It motivated Peano to give an abstract formulation of some of Grassmann's ideas in his *Geometric Calculus* (1888).

In the last chapter of this work, entitled "Transformations of linear systems," Peano gave an axiomatic definition of a vector space over the reals. He called it a "linear system." It was in the modern spirit of axiomatics, more or less as we have it today. He postulated the closure operations, associativity, distributivity, and the existence of a zero element. This was defined to have the property $0 \times a = 0$ for every element a in the vector space. He defined $a - b$ to mean $a + (-1)b$ and showed that $a - a = 0$ and $a + 0 = a$. Another of his axioms was that $a = b$ implies $a + c = b + c$ for every c. As examples of vector spaces he gave the real numbers, the complex numbers, vectors in the plane or in 3-space, the set of linear transformations from one vector space to another, and the set of polynomials in a single variable. See [23].

Peano also defined other concepts of linear algebra, including dimension and linear transformation, and proved a number of theorems. For example, he defined the dimension of a vector space as the maximum number of linearly independent elements (but did not prove that it is the same for every choice of such elements), and showed that any set of n linearly independent elements in an n-dimensional vector space constitutes a basis. He noted that if the set of polynomials in a single variable is at most of degree n, the dimension of the resulting vector space is $n + 1$, but if there is no restriction on the degree, the resulting vector space is infinite-dimensional. See [9], [23].

Peano's work was ignored for the most part, probably because axiomatics was in its infancy, and perhaps because the work was tied so closely to geometry, setting aside other important contexts of the ideas of linear algebra, which we have described above.

A word about the axiomatic method. Although it became well established only in the early decades of the twentieth century, it was "in the air" in the last two decades of the nineteenth. For example, there appeared axiomatic definitions of groups and fields, the positive integers (cf. the Peano axioms), and projective geometry.

In 1918, in his book *Space, Time, Matter*, which dealt with general relativity, Weyl axiomatized *finite-dimensional* real vector spaces, apparently unaware of Peano's work. The definition appears in the first chapter of the book, entitled *Foundations of Affine Geometry*. As in Peano's case, this was not quite the modern definition. It took time to get it just right! See [23].

In his doctoral dissertation of 1920 Banach axiomatized complete normed vector spaces (now called Banach spaces) over the reals. The first thirteen axioms are those of a vector space, in very much a modern spirit. He put it thus:

> I consider sets of elements about which I postulate certain properties; I deduce from them certain theorems, and I then prove for each particular functional domain that the postulates adopted are true for it.

In her 1921 paper "Ideal theory in rings" Emmy Noether introduced modules, and viewed vector spaces as special cases (see Chapter 6.2 and 6.3). We thus see vector spaces arising in three distinct contexts: geometry, analysis, and algebra. In his 1930 classic text *Modern Algebra* van der Waerden has a chapter entitled Linear Algebra [25]. Here for the first time the term is used in the sense in which we employ it today. Following in Noether's footsteps, he begins with the definition of a module over a (not necessarily commutative) ring. Only on the following page does he define a vector space!

References

1. I. G. Bashmakova and G. S. Smirnova, *The Beginnings and Evolution of Algebra*, The Mathematical Association of America, 2000. (Translated from the Russian by A. Shenitzer.)
2. N. Bourbaki, *Elements of the History of Mathematics*, Springer-Verlag, 1994.
3. T. Crilly, A gemstone in matrix algebra, *Math. Gazette* 1992, **76**: 182–188.
4. T. Crilly, Cayley's anticipation of a generalized Cayley-Hamilton theorem, *Hist. Math.* 1978, **5**: 211–219.
5. M. J. Crowe, *A History of Vector Analysis*, University of Notre Dame Press, 1967.
6. S. S. Demidov, On the history of the theory of linear differential equations, *Arch. Hist. Exact Sc.* 1983, **28**: 369–387.
7. J. Dieudoneé, *Abregé d'Histoire des Mathématiques 1700–1900*, vol. I, Hermann, 1978.
8. J.-L. Dorier (ed.), *On the Teaching of Linear Algebra*, Kluwer, 2000.
9. J.-L. Dorier, A general outline of the genesis of vector space theory, *Hist. Math.* 1985, **22**: 227–261.
10. D. Fearnley-Sander, Hermann Grassmann and the creation of linear algebra, *Amer. Math. Monthly* 1979, **86**: 809–817.
11. J. A. Goulet, The principal axis theorem, *The UMAP Journal* 1983, **4**: 135–156.
12. I. Grattan-Guinness and W. Ledermann, Matrix theory, in: *Companion Encyclopedia of the History and Philosophy of the Mathematical Sciences*, ed. by I. Grattan-Guinness, Routledge, 1994, vol. 1, pp. 775–786.
13. J. Gray, Finite-dimensional vector spaces, in: *Companion Encyclopedia of the History and Philosophy of the Mathematical Sciences*, ed. by I. Grattan-Guinness, Routledge, 1994, vol. 2, pp. 947–951.
14. T. L. Hankins, *Sir William Rowan Hamilton*, The Johns Hopkins University Press, 1980.
15. T. Hawkins, Weierstrass and the theory of matrices, *Arch. Hist. Exact Sc.* 1977, **17**: 119–163.
16. T. Hawkins, Another look at Cayley and the theory of matrices, *Arch. Int. d'Hist. des Sc.* 1977, **27**: 82–112.
17. T. Hawkins, Cauchy and the spectral theory of matrices, *Hist. Math.* 1975, **2**: 1–29.
18. T. Hawkins, The theory of matrices in the 19th century, *Proc. Int. Congr. Mathematicians* (Vancouver), vol. 2, 1974, pp. 561–570.
19. V. J. Katz, *A History of Mathematics*, 2nd ed., Addison-Wesley, 1998.
20. V. J. Katz, Historical ideas in teaching linear algebra, in *Learn from the Masters*, ed. by F. Swetz et al, Math. Assoc. of America, 1995, pp. 189–206.
21. E. Knobloch, Determinants, in: *Companion Encyclopedia of the History and Philosophy of the Mathematical Sciences*, ed. by I. Grattan-Guinness, Routledge, vol. 1, 1994, pp. 766–774.
22. E. Knobloch, From Gauss to Weierstrass: determinant theory and its historical evaluations, in C. Sasaki et al (eds), *Intersection of History and Mathematics*, Birkhäuser, 1994, pp. 51–66.
23. G. H. Moore, An axiomatization of linear algebra: 1875–1940, *Hist. Math.* 1995, **22**: 262–303.
24. T. Muir, The Theory of Determinants in the Historical Order of Development, 4 vols,, Dover, 1960. (The original work was published by Macmillan, 1890–1923.)
25. B. L. van der Waerden, *Modern Algebra*, 2 vols., Springer-Verlag, 1930–31.

6

Emmy Noether and the Advent of Abstract Algebra

The prominent algebraist Irving Kaplansky called Emmy Noether the "mother of modern algebra." The equally prominent Saunders MacLane asserted that "abstract algebra, as a conscious discipline, starts with Noether's 1921 paper "Ideal Theory in Rings." Hermann Weyl claimed that she "changed the face of algebra by her work." Let us attempt to do justice to these assertions.

According to van der Waerden, the essence of Noether's mathematical credo is contained in the following maxim:

> All relations between numbers, functions and operations become perspicuous, capable of generalization, and truly fruitful after being detached from specific examples, and traced back to conceptual connections.

We identify these ideas with the abstract, axiomatic approach in mathematics. They sound commonplace to us. But they were not so in Noether's time. In fact, they are commonplace today in considerable part because of her work.

Algebra in the nineteenth century was concrete by our standards. It was connected in one way or another with real or complex numbers. For example, some of the great contributors to algebra in the nineteenth century, mathematicians whose works shaped the algebra of the twentieth century, were Gauss, Galois, Jordan, Kronecker, Dedekind, and Hilbert. Their algebraic works dealt with quadratic forms, cyclotomy, field extensions, permutation groups, ideals in rings of integers of algebraic number fields, and invariant theory. All of these works were related in one way or another to real or complex numbers.

Moreover, even these important works in algebra were viewed in the nineteenth century, in the overall mathematical scheme, as secondary. The primary mathematical fields in that century were analysis (complex analysis, differential equations, real analysis), and geometry (projective, noneuclidean, differential, and algebraic). But after the work of Noether and others in the 1920s, algebra became central in mathematics.

It should be noted that Noether was not the only, nor even the only major contributor to the abstract, axiomatic approach in algebra. Among her predecessors who contributed to the genre were Cayley and Frobenius in group theory, Dedekind in lattice theory, Weber and Steinitz in field theory, and Wedderburn and Dickson in the theory of hypercomplex systems. Among her contemporaries, Albert in the U.S. and Artin in Germany stand out.

The "big bang" theory rarely applies when dealing with the origin of mathematical ideas. So also in Noether's case. The concepts she introduced and the results she established must be viewed against the background of late nineteenth-and early 20th century contributions to algebra. She was particularly influenced by the works of Dedekind. In discussing her contributions she frequently used to say, with characteristic modesty: "It can already be found in Dedekind's work" ("Es steht schon bei Dedekind"). In commenting on them, I will thus be considering their roots in Dedekind's work and in that of others from which she drew inspiration and on which she built.

Noether contributed to the following major areas of algebra: invariant theory (1907–1919), commutative algebra (1920–1929), noncommutative algebra and representation theory (1927–1933), and applications of noncommutative algebra to problems in commutative algebra (1932–1935). She thus dealt with just about the whole range of subject matter of the algebraic tradition of the nineteenth and early twentieth centuries (with the possible exception of group theory proper). What is significant is that she transformed that subject matter, thereby originating a new algebraic tradition—what has come to be known as modern or abstract algebra.

I will now discuss Noether's contributions to each of the above areas.

6.1 Invariant theory

Noether's statement (quoted above) that her ideas are already in Dedekind's work, could, with equal validity, have been put as "It all started with Gauss." Indeed, invariant theory dates back to Gauss' study of binary quadratic forms in his *Disquisitiones Arithmeticae*. Gauss defined an equivalence relation on such forms and showed that the discriminant is an invariant of the form under equivalence. See Chapter 3.2.

A second important source of invariant theory is projective geometry, which originated in the 1820s. A significant problem was to distinguish euclidean from projective properties of geometric figures. The projective properties turned out to be those invariant under "projective transformations."

Formally, invariant theory began with Cayley and Sylvester in the late 1840s. Cayley used it to bring to light the deeper connections between metric and projective geometry. Although important connections with geometry were maintained throughout the nineteenth and early twentieth centuries, invariant theory soon became an area of investigation independent of its relations to geometry. In fact, it became an important branch of *algebra* in the second half of the 19th century. To Sylvester:

> All algebraic inquiries, sooner or later, end at the Capitol of modern algebra over whose shining portal is inscribed the Theory of Invariants.

An important problem of the abstract theory of invariants was to discover invariants of various "forms." Many of the major mathematicians of the second half of the nineteenth century worked on the computation of invariants of specific forms. This led to the major problem of invariant theory, namely to determine a complete system of invariants—a basis—for a given form. That is, to find invariants of the form—it was conjectured that finitely many would do—such that every other invariant could be expressed as a combination of these

Cayley showed in 1856 that the finitely many invariants he had found earlier for binary quartic forms (i.e., forms of degree four in two variables) are a complete system. About ten years later Gordan proved that every binary form, of any degree, has a finite basis. Gordan's proof of this important result was computational—he exhibited a complete system of invariants. See Chapter 8.1.1 for further details.

In 1888 Hilbert astonished the mathematical world by announcing a new, conceptual approach to the problem of invariants. The idea was to consider, instead of invariants, expressions in a finite number of variables, in short, the polynomial ring in those variables. Hilbert then proved what came to be known as the *Basis Theorem*, namely that every ideal in the ring of polynomials in finitely many variables with coefficients in a field has a finite basis. A corollary was that every form, of any degree, in any number of variables, has a finite complete system of invariants.

Gordan's reaction to Hilbert's proof, which did not explicitly exhibit the complete system of invariants, was that "this is not mathematics; it is theology." When Hilbert later gave a constructive proof of this result (which he, however, did not consider significant), it elicited the following response from Gordan: "I have convinced myself that theology also has its advantages."

Noether's thesis, written under Gordan in 1907, was entitled "On Complete Systems of Invariants for Ternary Biquadratic Forms." The thesis was computational, in the style of Gordan's work. It ended with a table of the complete system of 331 invariants for such a form. She was later to describe her thesis as "a jungle of formulas."

She obtained, however, several notable results on invariants during the 1910s. First, using the methods she had developed in two papers (in 1915 and 1916) on the subject, she made a significant contribution to the problem, first posed by Dedekind, of finding a Galois extension of a given number field with a prescribed Galois group. Second, during her work in Göttingen on differential invariants, she used the calculus of variations to obtain the so-called Noether Theorem, still important in mathematical physics. The physicist Fez Gursey said of this contribution:

> The key to the relation of symmetry laws to conservation laws in physics is Emmy Noether's celebrated theorem which states that a dynamical system described by an action under a Lie group with n parameters admits n invariants (conserved quantities) that remain constant in time during the evolution of the system [15].

Alexandrov summarized her work on invariants by noting that it "would have been enough ... to earn her the reputation of a first class mathematician."

What was the route that led Emmy Noether from the computational theory of invariants to the abstract theory of rings and modules? "A greater contrast," noted Weyl, "is hardly imaginable than between her first paper, the dissertation, and her works of maturity."

In 1910 Gordan retired from the University of Erlangen and was soon replaced by Ernst Fischer. He too was a specialist in invariant theory, but invariant theory of the Hilbert persuasion. Noether came under his influence and gradually made the change from Gordan's algorithmic approach to invariant theory to Hilbert's conceptual approach.

Later work on invariants brought her in contact with the famous joint paper of Dedekind and Weber on the arithmetic theory of algebraic functions (see Chapter 3.2.2). She became "sold" on Dedekind's approach and ideas, and this determined the direction of her future work.

6.2 Commutative algebra

The two major sources of commutative algebra are algebraic geometry and algebraic number theory. Emmy Noether's two seminal papers of 1921 and 1927 on the subject can be traced, respectively, to these two sources. In these papers, entitled, respectively, "Ideal Theory in Rings" and "Abstract Development of Ideal Theory in Algebraic Number Fields and Function Fields," she broke fundamentally new ground, originating "a new and epoch-making style of thinking in algebra" [31].

Algebraic geometry had its origins in the study, begun in the early nineteenth century, of abelian functions and their integrals. This analytic approach to the subject gradually gave way to geometric, algebraic, and arithmetic means of attack. In the algebraic context, the main object of study is the ring of polynomials $k[x_1, x_2, \ldots, x_n]$, k a field (in the nineteenth century k was the field of real or complex numbers). Hilbert in the nineteenth century, and Lasker and Macauley in the early twentieth century, had shown that in such a ring every ideal is a finite intersection of primary ideals, with certain uniqueness properties. (Geometrically, the result says that every variety is a unique, finite union of irreducible varieties.) See Chapter 3.2.2 for details.

In her 1921 paper Noether generalized this result to arbitrary commutative rings with the ascending chain condition (a.c.c.). Her main result was that in such a ring every ideal is a finite intersection of primary ideals, with accompanying uniqueness properties.

What was so significant about this paper which (we recall) MacLane singled out as marking the beginning of abstract algebra as a conscious discipline?

First and foremost was the isolation of the a.c.c. as the crucial concept needed in the proof of the main result. In fact, the proof "rested entirely on elementary consequences of the chain condition and ... [was] startling in ... simplicity" [15]. Earlier proofs of the corresponding result for polynomial rings involved considerable computation, such as elimination theory and the geometry of algebraic sets.

The a.c.c. did not originate with Noether. Dedekind (in 1894) and Lasker (in 1905) used it, but in concrete settings of rings of algebraic integers and of polynomials, respectively. Moreover, the a.c.c. was for them incidental rather than of major consequence. Noether's isolation of the a.c.c. as an important concept was a watershed. Thanks to her work, rings with the a.c.c., now called *Noetherian rings*, have been singled out for special attention. In fact, commutative algebra has been described as the study of commutative Noetherian rings. As such, the subject had its formal genesis in Noether's 1921 paper.

Another fundamental concept which she highlighted in the 1921 paper was that of a ring. This concept, too, did not originate with her. Dedekind (in 1871) introduced it as a subset of the complex numbers closed under addition, subtraction, and multiplication, and called it an "order." Hilbert, in his famous Report on Number Theory (Zahlbericht) of 1897, coined the term "ring," but only in the context of rings of integers of algebraic number fields. Fraenkel (in 1914) gave essentially the modern definition of ring, but postulated two extraneous conditions. Noether in her 1921 paper gave the definition in current use (given also by Sono in 1917, but this went unnoticed). See Chapter 3.3.

But it was not merely Noether's *definition* of the concept of a ring which proved important. Through her groundbreaking papers in which that concept played an essential role, and of which the 1921 paper was an important first, she brought it into prominence as a central notion of algebra. It immediately began to serve as the starting point for much of abstract algebra, taking its rightful place alongside the concepts of group and field, already reasonably well established at that time.

Noether also began to develop in the 1921 paper a general theory of ideals for commutative rings. Notions of prime, primary, and irreducible ideal, of intersection and product of ideals, of congruence modulo an ideal—in short, much of the machinery of ideal theory, appears here.

Toward the end of the paper she defined the concept of *module* over a noncommutative ring and showed that some of the earlier decomposition results for ideals carry over to submodules. We will discuss modules in connection with her work in noncommutative algebra.

To summarize, the 1921 paper introduced and gave prominence to what came to be some of the basic concepts of abstract algebra, namely ring, module, ideal, and the a.c.c. Beyond that, it introduced, and began to show the efficacy of, a new way of doing algebra—abstract, axiomatic, conceptual. No mean accomplishment for a single paper!

Noether's 1927 paper had its roots in algebraic number theory and, to a lesser extent, in algebraic geometry. The sources of algebraic number theory are Gauss' theory of quadratic forms of 1801, his study of biquadratic reciprocity of 1832 (in which he introduced the Gaussian integers), and attempts in the early nineteenth century to prove Fermat's Last Theorem. In all cases the central issue turned out to be unique factorization in rings of integers of algebraic number fields. When examples of such rings were found in which unique factorization fails, the problem became to try to "restore," in some sense, the "paradise lost." This was achieved by Dedekind in 1871 (and, in a different way, by Kronecker in 1882) when he showed that unique factorization can be reestablished if one considers factorization of ideals, which he had introduced for this purpose, rather than of elements. His main result was that if R is the ring of integers of an algebraic number field, then every ideal of R is a unique product of prime ideals. See Chapter 3.2.1 for details.

Riemann introduced "Riemann surfaces" in the 1850s in order to facilitate the study of (multivalued) algebraic functions. His methods were, however, nonrigorous, and depended on physical considerations. In 1882 Dedekind and Weber wrote an all-important paper whose aim was to give rigorous, algebraic expression to some of Riemann's ideas on complex function theory, in particular to his notion of a Riemann surface. Their idea was to establish an analogy between algebraic number fields and algebraic function fields, and to carry over the machinery and results of the former to the latter. They succeeded admirably, giving (among other things) a purely algebraic definition of a Riemann surface, and an algebraic proof of the fundamental Riemann–Roch Theorem. At least as importantly, they pointed to what proved to be a most fruitful idea, namely the interplay between algebraic number theory and algebraic geometry.

More specifically, just as in algebraic number theory one associates an algebraic number field $Q(u)$ with a given algebraic number u, so in algebraic geometry one associates an algebraic function field $\mathbf{C}(x, y)$ with a given algebraic function. $\mathbf{C}(x, y)$ consists of polynomials in x and y with complex coefficients, where y satisfies a polynomial equation with coefficients in $\mathbf{C}(x)$ (i.e., y is algebraic over $\mathbf{C}(x)$). If A is the "ring of integers" of $\mathbf{C}(x, y)$, that is, A consists of the roots in $\mathbf{C}(x, y)$ of monic polynomials with coefficients in $\mathbf{C}[x]$, then a major result of the Dedekind–Weber paper is that every ideal in A is a unique product of prime ideals. See Chapter 3.2.2.

In her 1927 paper Noether generalized the above decomposition results for algebraic number fields and function fields to commutative rings. In fact, she characterized those commutative rings in which every ideal is a unique product of prime ideals. Such rings are now called *Dedekind domains*. She showed that R is a Dedekind domain if and only if

(1) R satisfies the a.c.c.,
(2) R/I satisfies the d.c.c. for every nonzero ideal I of R,
(3) R is an integral domain, and
(4) R is integrally closed in its field of quotients.

Condition (4) proved particularly significant since it singled out the basic notion of integral dependence (related to that of integral closure). This concept, already present in Dedekind's work on algebraic numbers, has proved to be of fundamental importance in commutative algebra. As Gilmer notes, "the concept of integral dependence is to *Aufbau* [Noether's 1927 paper] what the a.c.c. is to *Idealtheorie* [her 1921 paper]" [12].

Among other basic results she proved in this paper are:

(a) the (by now standard) isomorphism and homomorphism theorems for rings and modules,
(b) that a module M has a composition series if and only if it satisfies both the a.c.c. and d.c.c., and

(c) that if an R-module M is finitely generated and R satisfies the a.c.c. (d.c.c.), then so does M.

To summarize Noether's contributions to commutative algebra: In addition to proving important results, she introduced concepts and developed techniques which have become standard tools of the subject. In fact, her 1921 and 1927 papers, combined with those of Krull of the 1920s, are said to have created the subject of commutative algebra.

6.3 Noncommutative algebra and representation theory

Before her ideas in commutative algebra had been fully assimilated by her contemporaries, Noether turned her attention to the other major algebraic subjects of the nineteenth and early twentieth centuries, namely hypercomplex number systems (what we now call associative algebras) and groups (in particular group representations). She extended and unified these two subjects through her abstract, conceptual approach, in which module-theoretic ideas that she had used in the commutative case played a crucial role.

The theory of hypercomplex systems began with Hamilton's 1843 introduction of the quaternions. At the end of the nineteenth century, E. Cartan, Frobenius, and Molien gave structure theorems for such systems over the real and complex numbers, and in 1907 Wedderburn extended these to hypercomplex systems over arbitrary fields. In the spirit of Noether's work in commutative algebra, Artin extended Wedderburn's results to noncommutative, semi-simple rings with the descending chain condition. See Chapter 3.1 and 3.4 for details.

Groups were the first of the algebraic systems to be developed extensively. By the end of the nineteenth century they began to be studied abstractly (see Chapter 2). An important tool in that study was representation theory, developed by Burnside, Frobenius, and Molien in the 1890s. The idea was to study, instead of the abstract group, its concrete representations in terms of matrices. (A representation of a group is a homomorphism of the group into the group of invertible matrices of some given order.)

In her 1929 paper "Hypercomplex numbers and representation theory" Noether framed group representation theory in terms of the structure theory of hypercomplex systems. The main tool in this approach was the *module*. The idea was to associate with each representation φ of G by invertible matrices with entries in some field k, a $k(G)$-module V called the *representation module* of φ ($k(G)$ is the *group algebra* of G over k).

Conversely, any $k(G)$-module M gives rise to a representation ψ of G. This establishes a one-one correspondence between representations of G over k and $k(G)$-modules. The standard concepts of representation theory can now be phrased in terms of modules. For example, two representations are equivalent if and only if their representation modules are isomorphic; a representation is irreducible if and only if its representation module is simple. The techniques of module theory, and the structure theory of hypercomplex systems (applied to the hypercomplex system $k(G)$) could now be used to recast the foundations of group representation theory. See [8] and [18] for details.

Noether's work in this area created a very effective conceptual framework in which to study representation theory. For example, while the classical (computational) approach to representation theory is valid only over the field of complex numbers, or at best over an algebraically closed field of characteristic 0, Noether's approach remains meaningful for any field, of any characteristic.

The use of arbitrary fields in representation theory became important in the 1930s when Brauer began his pioneering studies of *modular representations*, that is, those in which the characteristic of the field divides the order of the group. Noether's ideas also "planted the seed of modern integral representation theory" [18], that is, representation theory over commutative rings rather than over fields. She herself extended the representation theory of groups to that of semi-simple Artinian rings; here she needed the concept of a *bimodule*.

A word about modules, which were so central in Noether's work in both commutative and noncommutative algebra. Dedekind, in connection with his 1871 work in algebraic number theory, was the first to use the term "module," but to him it meant a subgroup of the additive group of complex numbers (i.e., a Z-module). In 1894 he developed an extensive theory of such modules. Lasker, in his 1905 work on decomposition of polynomial rings, used the terms "module" and "ideal" interchangeably (the former he applied to polynomial rings over **C**, the latter to such rings over Z).

Noether was the first to use the notion of module abstractly, with a ring as domain of operators, and to recognize its potential. In fact, it is through her work that the concept of module became the central concept of algebra that it is today. Indeed, modules are important not only because of their unifying, but also because of their linearizing, power. They are, after all, generalizations of vector spaces, and many of the standard vector-space constructions, such as subspace, quotient space, direct sum, and tensor product carry over to modules. (We know of the power of linearization in analysis. Modules can be said to provide analogous power in algebra.)

The importance of the invention of homological algebra was that it carried the process of linearization far forward by developing tools for its implementation. For example, the functors "Ext" and "Tor" measure the extent to which modules over general rings "misbehave" when compared to modules over fields, namely vector spaces.

6.4 Applications of noncommutative to commutative algebra

Noether believed that the theory of noncommutative algebra is governed by simpler laws than that of commutative algebra. In her 1932 plenary address at the International Congress of Mathematicians in Zurich, entitled Hypercomplex Systems and their Relations to Commutative Algebra and Number Theory, she outlined a program putting that belief into practice. Her program has been called "a foreshadowing of modern cohomology theory" [25]. The ideas on factor sets contained therein were soon used by Hasse and Chevalley "to obtain some of the main results on global and local class field theory" [15]. Noether's own immediate objective was to apply the theory of central simple algebras, as developed by her, Brauer, and others, to problems in class field theory.

Some of Noether's ideas (and those of others) on the interplay between commutative and noncommutative algebra had born fruit with the proof of the celebrated Albert–Brauer–Hasse–Noether Theorem. This result, called by Jacobson "one of the high points of the theory of algebras," gives a complete description of finite-dimensional division algebras over algebraic number fields. It is important in the study of finite-dimensional algebras and of group representations.

To bring out the context of the above theorem, it should be noted that Wedderburn's 1907 structure theorems for finite-dimensional algebras reduced their study to that of nilpotent algebras and division algebras (see Chapter 3.1). Since the unravelling of the structure of the former seemed (and still seems, despite considerable progress) "hopeless," attention focussed on the latter.

Considerable progress on the structure of division algebras was made in the late 1920s and early 1930s. The Albert–Brauer–Hasse–Noether Theorem was a high point of these researches. It should be stressed, however, that even today much is still unknown about finite-dimensional division algebras.

6.5 Noether's legacy

The concepts she introduced, the results she obtained, and the mode of thinking she promoted, have become part of our mathematical culture. As Alexandrov put it:

> It was she who taught us to think in terms of simple and general algebraic concepts—homomorphic mappings, groups and rings with operators, ideals—and not in cumbersome algebraic computations; and [she] thereby opened up the path to finding algebraic principles in places where such principles had been obscured by some complicated special situation . . . [1].

Moreover, as Weyl noted:

> Her significance for algebra cannot be read entirely from her own papers; she had great stimulating power and many of her suggestions took shape only in the works of her pupils or co-workers [31].

Indeed, Weyl himself acknowledged his indebtedness to her in his work on groups and quantum mechanics. Among others who have explicitly mentioned her influence on their algebraic works are such prominent algebraists as Artin, Deuring, Hasse, Jacobson, Krull, and Kurosh.

Another important vehicle for the spread of Noether's ideas was the now classic treatise of van der Waerden entitled *Modern Algebra*, first published in 1930. It was based on lectures of Noether and Artin. Its wealth of beautiful and powerful ideas, brilliantly presented by van der Waerden, has nurtured a generation of mathematicians. The book's immediate impact is poignantly described by Dieudonné and G. Birkhoff, respectively:

> I was working on my thesis at that time; it was 1930 and I was in Berlin. I still remember the day that van der Waerden came out on sale. My ignorance in

algebra was such that nowadays I would be refused admittance to a university. I rushed to those volumes and was stupefied to see the new world which opened before me. At that time my knowledge of algebra went no further than *mathematiques spéciales*, determinants, and a little on the solvability of equations and unicursal curves. I had graduated from the École Normale and I did not know what an ideal was, and only just knew what a group was! This gives you an idea of what a young French mathematician knew in 1930 [10].

Even in 1929, its concepts and methods [i.e., those of "modern algebra"] were still considered to have marginal interest as compared with those of analysis in most universities, including Harvard. By exhibiting their mathematical and philosophical unity and by showing their power as developed by Emmy Noether and her other younger colleagues (most notably E. Artin, R. Brauer, and H. Hasse), van der Waerden made "modern algebra" suddenly seem central in mathematics. It is not too much to say that the freshness and enthusiasm of his exposition electrified the mathematical world—especially mathematicians under 30 like myself [3].

A number of mathematicians and historians of mathematics have spoken of the "algebraization of mathematics" in the twentieth century. Witness the terminological penetration of algebra into such fields as algebraic geometry, algebraic topology, algebraic number theory, algebraic logic, topological algebra, Banach algebras, von Neumann algebras, Lie groups, and normed rings. Noether's influence is evident directly in several of these fields and indirectly in others. She too seemed to have acknowledged that, when she said in a letter to Hasse in 1931:

My methods are really methods of working and thinking; this is why they have crept in everywhere anonymously [9].

Alexandrov and Hopf confirm this in the preface to their book on topology:

Emmy Noether's general mathematical insights were not confined to her specialty—algebra—but affected anyone who came in touch with her [9].

In fact, they too, and more importantly, algebraic topology, were major beneficiaries of her insights. Jacobson confirms this:

As is quite well known, it was Emmy Noether who persuaded Alexandrov and ... Hopf to introduce group theory into combinatorial topology and to formulate the then existing simplicial homology theory in group theoretic terms in place of the more concrete setting of incidence matrices [15].

Algebraic geometry is another area which witnessed very extensive algebraization beginning in the late 1920s and early 1930s. The testimonies of Zariski and van der Waerden, respectively, two of its foremost practitioners, who were deeply involved in this process of algebraization, are revealing:

It was a pity that my Italian teachers never told me there was such a tremendous development of the algebra which is connected with algebraic geometry. I only discovered this much later, when I came to the United States [23].

When I came to Göttingen in 1924, a new world opened up before me. I learned from Emmy Noether that the tools by which my questions [in algebraic geometry] could be handled had already been developed ... [24].

Noether was a visiting professor in Moscow in 1928–1929. Alexandrov described the impact she has had on Pontryagin's work in the theory of continuous groups (topological algebra):

It is not hard to follow the influence of Emmy Noether on the developing mathematical talent of Pontryagin; the strong algebraic flavour in Pontryagin's work undoubtedly profited greatly from his association with Emmy Noether [1].

I will give the last word to Garrett Birkhoff who, in an article in 1976 describing the rise of abstract algebra from 1936 to 1950, said the following:

If Emmy Noether could have been at the 1950 [International] Congress [of Mathematicians], she would have felt very proud. Her concept of algebra had become central in contemporary mathematics. And it has continued to inspire algebraists ever since [4].

References

1. P. S. Alexandrov, In memory of Emmy Noether, in *Emmy Noether, 1882–1935*, by A. Dick, Birkhäuser, 1981, pp. 153–179.
2. M. F. Atiyah and I. G. Macdonald, *Introduction to Commutative Algebra*, Addison Wesley, 1969.
3. G. Birkhoff, Current trends in algebra, *Amer. Math. Monthly* 1973, **80**: 760–782.
4. G. Birkhoff, (a) The rise of modern algebra to 1936 and (b) The rise of modern algebra, 1936 to 1950, in *Men and Institutions in American Mathematics*, ed. by D. Tarwater et al, Texas Tech Press, 1976, pp. 41–63 and 65–85.
5. N. Bourbaki, Historical note, in his *Commutative Algebra*, Addison-Wesley, 1972, pp. 579–606.
6. J. W. Brewer and M. K. Smith (eds), *Emmy Noether: A Tribute to her Life and Work*, Marcel Dekker, 1981.
7. T. Crilly, (a) The rise of Cayley's invariant theory (1841–1862), and (b) The decline of Cayley's invariant theory (1863–1895), *Hist. Math.* 1986 **13**: 241–254, and 1988, **15**: 332–347.
8. C. W. Curtis and I. Reiner, *Representation Theory of Finite Groups and Associative Algebras*, Wiley, 1962.
9. A. Dick, *Emmy Noether, 1882–1935*, Birkhäuser, 1981.
10. J. Dieudonné, The work of Nicolas Bourbaki, *Amer. Math. Monthly* 1970, **77**: 134–145.
11. P. Dubreil, Emmy Noether, *Cahiers du Sém. d'Hist. des Math.* 1986, **7**: 15–27.
12. R. Gilmer, Commutative ring theory, in *Emmy Noether: A Tribute to her Life and Work*, ed. by J. W. Brewer and M. K. Smith, Marcel Dekker, 1981, pp. 131–143.
13. T. Hawkins, Hypercomplex numbers, Lie groups, and the creation of group representation theory, *Arch. Hist. Ex. Sc.* 1976/77 **16**: 17–36.
14. N. Jacobson, *Basic Algebra*, 2 vols., Freeman, 1974 and 1980.

15. N. Jacobson (ed.), *Emmy Noether: Collected Papers*, Springer-Verlag, 1983. Contains an introduction by Jacobson to Noether's works.
16. I. Kaplansky, Commutative rings, in *Proc. of Conf. on Commutative Algebra*, ed. by J. W. Brewer and E. A. Rutter, Springer-Verlag, 1973, pp. 153–166.
17. C. H. Kimberling, Emmy Noether and her influence, in *Emmy Noether: A Tribute to her Life and Work*, ed. by J. W. Brewer and M. K. Smith, Marcel Dekker, 1981, pp. 3–61.
18. T. Y. Lam, Representation theory, in *Emmy Noether: A Tribute to her Life and Work*, ed. by J. W. Brewer and M. K. Smith, Marcel Dekker, 1981, pp. 145–156.
19. S. MacLane, History of abstract algebra, in *American Mathematical Heritage: Algebra and Applied Mathematics*, ed. by D. Tarwater et al, Texas Tech Press, 1981, pp. 3–35.
20. S. MacLane, Mathematics at the University of Göttingen (1931–1933), in *Emmy Noether: A Tribute to her Life and Work*, ed. by J. W. Brewer and M. K. Smith, Marcel Dekker, 1981, pp. 65–78.
21. U. Merzbach, Historical contexts, in *Emmy Noether in Bryn Mawr*, ed. by B. Srinivasan and J. Sally, Springer-Verlag, 1983, pp. 161–171.
22. A. F. Monna, L'algébrisation de la mathématique, réflexions historiques, *Comm. Math. Inst., Rijksuniversiteit*, Utrecht, 1977.
23. C. Parikh, *The Unreal Life of Oscar Zariski*, Academic Press, 1991.
24. H. Pollard and H. G. Diamond, *The Theory of Algebraic Numbers*, Math. Assoc. of America, 1975.
25. M. K. Smith, Emmy Noether's contributions to mathematics, Unpublished notes (13 pp, ca 1976).
26. B. Srinivasan and J. Sally (eds.), *Emmy Noether in Bryn Mawr*, Springer-Verlag, 1983.
27. B. L. van der Waerden, Obituary of Emmy Noether, in *Emmy Noether, 1882–1935*, by A. Dick, Birkhäuser, 1981, pp. 100–111.
28. B. L. van der Waerden, The foundations of algebraic geometry from Severi to André Weil, *Arch. Hist. Ex. Sc.* 1970–71, **7**: 171–180.
29. B. L. van der Waerden, On the sources of my book *Moderne Algebra*, *Hist. Math.* 1975, **2**: 31–40.
30. B. L. van der Waerden, The school of Hilbert and Emmy Noether, *Bull. Lond. Math. Soc.* 1983, **15**: 1–7.
31. H. Weyl, Memorial address, in *Emmy Noether, 1882–1935*, by A. Dick, Birkhäuser, 1981, pp. 112–152.

A Course in Abstract Algebra
Inspired by History

I propose to describe a course in abstract algebra which I taught in an In-Service Programme for Teachers of Mathematics at our university. Students do not follow this course with another in abstract algebra. This presented an opportunity and a challenge: what are some of the major ideas of abstract algebra that I would want to impart to students? What algebraic legacy would I want to leave them with? Since the students were high-school teachers of mathematics, I wanted the course also to have at least broad relevance to their concerns as teachers. Let me add that I believe this course can be readily adapted to other student audiences.

The above remarks suggested (to me, at least) that the history of mathematics should play an important role in the course. History points to the sources of abstract algebra, hence to some of its central ideas; it provides motivation; and it makes the subject come to life.

To set the context for the course, here is a history of abstract algebra—in 100 words or less.

Prior to the nineteenth century algebra meant essentially the study of polynomial equations. In the twentieth century algebra became the study of abstract, axiomatic systems such as groups, rings, and fields. The transition from the so-called classical algebra of polynomial equations to the so-called modern algebra of axiom systems occurred in the nineteenth century. Modern algebra came into existence principally because mathematicians were unable to solve classical problems by classical (pre-nineteenth century) means. They invented the concepts of group, ring, and field to help them solve such problems. The previous chapters have, I trust, made this point.

The upshot of this mini-history of algebra is to help focus on the major theme of the course, namely showing how abstract algebra originated in, and sheds light on, the solution of "concrete" problems. It is a confirmation of Whitehead's dictum that "the utmost abstractions are the true weapons with which to control our thought of concrete fact." What I do in the course can be represented schematically as follows:

$$\text{Problems} \rightarrow \text{Abstractions} \left< \begin{array}{l} \text{Solutions of original problems} \\ \\ \text{Solutions of other problems} \end{array} \right.$$

The item "Solutions of other problems" is intended to convey an important idea, namely that the abstract concepts whose introduction was motivated by concrete problems often superseded in importance the original problems which inspired them. In particular, the emerging new concepts and results were employed in the solution of other problems, often unrelated to, and sometimes more important than, the original problems which gave them birth. I will call the solutions of such problems "payoffs." Now to the problems.

Problem I: Why is $(-1)(-1) = 1$?

This problem is an instance of the issue of justification of the laws of arithmetic. It deals with relations between arithmetic and abstract algebra, and it gives rise to the concepts of ring, integral domain, ordered structure, and axiomatics.

The above problem became pressing for English mathematicians of the early nineteenth century, who wanted to set algebra—to them this meant the laws of operation with numbers—on an equal footing with geometry by providing it with logical justification. The task was tackled by members of the Analytical Society at Cambridge [19]. We will focus on Peacock's work, *Treatise of Algebra* (1830), which proved the most influential.

Peacock's major idea was to distinguish between "arithmetical algebra" and "symbolical algebra." The former referred to operations involving only *positive* numbers, hence in Peacock's view required no justification. For example, $a - (b-c) = a+c-b$ is a law of arithmetical algebra when $b > c$ and $a > (b - c)$. It becomes a law of symbolical algebra if no restrictions are placed on a, b, and c. In fact, *no interpretation of the symbols is called for*. Thus *symbolical algebra* is the subject—newly founded by Peacock—of operations with symbols which need not refer to specific objects but which obey the laws of arithmetical algebra. Peacock's justification for identifying the laws of symbolical algebra with those of arithmetical algebra was his *Principle of Permanence of Equivalent Forms*, a type of Principle of Continuity going back at least to Leibniz:

> Whatever algebraic forms are equivalent when the symbols are general in form but specific in value, will be equivalent when the symbols are general in value as well as in form.

Thus Peacock *decreed* that the laws of arithmetic shall also be the laws of (symbolical) algebra—an idea not at all unlike the axiomatic approach to arithmetic. For example, we can use Peacock's *Principle* to prove that $(-x)(-y) = xy$:

Since $(a - b)(c - d) = ac + bd - ad - bc$ whenever $a > b$ and $c > d$, this being a law of arithmetic and hence requiring no justification, it also becomes a law of symbolical algebra—that is, without restrictions on a, b, c, and d. Letting $a = 0$ and $c = 0$ yields $(-b)(-d) = bd$.

The significance of Peacock's work was that symbols took on a life of their own, becoming objects of study in their own right rather than a language to represent relationships among numbers. Some have said that these developments signalled the birth of abstract algebra. See Chapter 1.8.

We now make a seventy-year leap forward and take a modern, Hilbertian approach to the above topic. The idea is to define (characterize) the integers axiomatically as an ordered integral domain in which the positive elements are well ordered ([17], [22]), just as Hilbert (in 1900) characterized the reals axiomatically as the maximal Archimedean ordered field [3], [11]. Of course, in the process we must define the various algebraic concepts that enter into the above characterization of the integers. We can then readily prove such laws as $(-a)(-b) = ab$ and $a \times 0 = 0$. This was done in the more general context of rings by Fraenkel in 1914. See Chapter 3.3.

Payoffs:
The following issues arise from the above account:

(a) How can we establish (prove) a law such as $(-1)(-1) = 1$? This question leads to *axioms*. We cannot prove everything.
(b) What axioms should we set down to give a description of the integers? This question enables us to introduce the concepts of ring, integral domain, ordered ring, and well ordering (induction).
(c) How do we know when we have enough axioms? Here we introduce the idea of *completeness* of a set of axioms.
(d) What does it mean to characterize the integers? This sets the stage for the introduction of the notion of *isomorphism*.
(e) Could we have used fewer axioms to characterize the integers? For example, $a + b = b + a$ is not needed. Here we come face to face with the concept of *independence* of a set of axioms.
(f) Are we at liberty to pick and choose axioms as we please? This question permits us to introduce the notion of *consistency*, and more broadly, the issue of freedom of choice in mathematics.

The innocent-looking problem $(-1)(-1) = 1$ can be a rich source of ideas!

Problem II: What are the integer solutions of $x^2 + 2 = y^3$?

This diophantine equation is an example of the famous Bachet equation $x^2 + k = y^3$, introduced in the seventeenth century and solved (theoretically) for arbitrary k only recently. The problem deals with relations between number theory and abstract algebra, and it gives rise to the concepts of *unique factorization domain* (UFD) and *euclidean domain*—important examples of commutative rings.

We begin with a simpler problem, namely to solve the diophantine equation $x^2 + y^2 = z^2$, with $(x, y) = 1$, that is, to find all primitive Pythagorean triples. Although the solution was known in ancient Greece over 2000 years ago, if not earlier, we are interested in an "algebraic" solution—a legacy of the nineteenth century.

The key idea is to factor the left side of $x^2 + y^2 = z^2$ and thus obtain the equation $(x+yi)(x-yi) = z^2$ in the domain $G = \{a+bi : a, b \in Z\}$ of Gaussian integers. This domain shares with the integers the property of unique factorization, as was shown by Gauss. In particular, since $x + yi$ and $x - yi$ are relatively prime in G (this follows because x and y are relatively prime in Z) and their product is a square, each is a square in G (this holds in any UFD). Thus $x + yi = (a + bi)^2 = (a^2 - b^2) + 2abi$. Comparing real and imaginary parts yields $x = a^2 - b^2$, $y = 2ab$, and since $x^2 + y^2 = z^2$, $z = a^2 + b^2$. Conversely, it is easily shown that for any $a, b \in Z$, $(a^2 - b^2, 2ab, a^2 + b^2)$ is a solution of $x^2 + y^2 = z^2$. We thus get all Pythagorean triples. It is easy to single out the primitive ones among them.

Coming back to $x^2 + 2 = y^3$, we proceed analogously by factoring the left side and get $(x + \sqrt{2}i)(x - \sqrt{2}i) = y^3$, an equation in the domain $D = \{a + b\sqrt{2}i : a, b \in Z\}$. Here too we can show that $(x + \sqrt{2}i, x - \sqrt{2}i) = 1$, hence $x + \sqrt{2}i$ and $x - \sqrt{2}i$ are cubes in D. In particular, $x + \sqrt{2}i = (a + b\sqrt{2}i)^3$. Simple algebra yields $x = \pm 5$, $y = 3$. Of course it is easy to see that these are solutions of $x^2 + 2 = y^3$. What the above shows is that they are the *only* solutions. This is essentially how Euler solved the equation. Of course one must show that D is a UFD.

The Fermat equation $x^3 + y^3 = z^3$ can be dealt with similarly: $z^3 = x^3 + y^3 = (x + y)(x + y\omega)(x + y\omega^2)$—an equation in the domain $E = \{a + b\omega : a, b \in Z, \omega$ a primitive cube root of $1\}$. The technical details are more complex here [1], [9].

Justifying the "details" in the solutions of the above three diophantine equations involves considerable work. In particular, we need to introduce the notions of unique factorization domain and euclidean domain and to discuss some of their arithmetic properties. The three diophantine equations can be solved in the indicated manner *because* the respective domains G, D, and E in which they were embedded are UFDs.

For historical details on the above see Chapter 3.2.

Payoffs:

(a) We can solve Fermat's problem about the representability of integers as sums of two squares by a careful scrutiny of the primes in the domain G of Gaussian integers [9], [18].

(b) In arithmetic domains in which unique factorization fails we introduce, following Dedekind, ideals. We can thereby obtain a proof of Fermat's Last Theorem—the unsolvability in nonzero integers of $x^p + y^p = z^p$—for all $p < 100$ [18]. Here appear the elements of a rich subject—algebraic number theory, in which the ideas of abstract algebra play a fundamental role. The subject originated to a large extent in attempts to solve such diophantine equations as we have considered above, in particular the Fermat equation. See Chapter 3.2.

Problem III: Can we trisect a 60° angle using only straightedge and compass?

This is an instance of one of the three famous classical construction problems going back to Greek antiquity. It deals with relations between geometry and abstract algebra, and it gives rise to the concepts of field and vector space. This is a standard problem, usually given following the presentation of Galois theory. I put it centre-stage as a means of providing a "gentle" introduction to fields.

The problem of trisection was posed about 2500 years ago but solved only in 1837, by Wantzel, following the introduction of the requisite algebraic machinery. One must persevere!

The initial key idea was the translation of the geometric problem into the language of classical algebra—numbers and equations. This occurred in the seventeenth century. Thus the basic goal became the construction of *real numbers*, often as roots of equations. ("Construction" will henceforth mean "construction with straightedge and compass.") How do fields and vector spaces enter the picture?

If a and b are constructible, so are $a + b, a - b, ab$, and $a/b(b \neq 0)$—all this is easy to show. Thus the constructible numbers form a field. But what are they?

Given a unit length 1, the above implies that we can construct all rational numbers Q. We can also construct, for example, $\sqrt{2}$, as the diagonal of a unit square. More generally, if a is constructible, so is $\sqrt{a}$. We can therefore construct the field $Q(\sqrt{a}) = \{p + q\sqrt{a} : p, q \in Q\}$. This introduces the important notion of field adjunction. The objective is to show that all constructible numbers can be obtained by an iteration of the adjunction of square roots.

To proceed we need a numerical measure of how far $Q(\sqrt{a})$ is removed from Q. This leads to the concept of degree of a field extension, here the dimension of $Q(\sqrt{a})$ as a vector space over Q. The problem of trisection is next phrased in terms of fields. This is now late-nineteenth-century abstract algebra. Enough machinery of field extensions is introduced—and not much more than that—to solve the trisection problem [12].

A word about history versus genesis. Wantzel solved the trisection problem in 1837, essentially as we do: he reduced the problem to the solution of polynomial equations, introduced irreducible polynomials and rational functions of a given number of elements, and derived conditions for constructibility in terms of the iteration of solutions of polynomial equations [28]. Although Wantzel's approach is similar in spirit to the modern one, he used neither fields nor vector spaces. We use both. Our approach in this course is genetic rather than strictly historical when this serves our purpose.

Payoffs:
(a) A characterization of the real numbers as a complete ordered field [3].
(b) A discussion of algebraic and transcendental numbers [8], [18].
(c) A characterization of finite fields [10], [17].
(d) Proof of a special case of Dirichlet's theorem on primes in arithmetic progression, namely that the sequence $1, 1 + b, 1 + 2b, 1 + 3b, \ldots, b$ an arbitrary positive integer, contains infinitely many primes. For this we need cyclotomic field extensions [8].

Problem IV: Can we solve $x^5 - 6x + 3 = 0$ by radicals?

Problems such as this, dealing with the solution of equations by radicals, gave rise to Galois theory (see Chapter 2). They touch on the relations between classical and abstract algebra.

Galois theory, in its modern incarnation, is a grand symphony on two major themes—groups and fields, and two minor themes—rings and vector spaces. Galois theory is thus a highlight of any course in abstract algebra. But to do it in detail would take most of an entire term. Moreover, the proofs of theorems are often rather long, and sometimes tedious, and the payoff is long in coming. The intent in this course, then, is to get across some of the central ideas of Galois theory—for example, the correspondence between groups and fields and what it is good for—often with examples rather than proofs.

We begin where the history of the subject begins—with Lagrange. He analyzed past solutions of the cubic and quartic to see if he could find in them a common method extendable to the quintic. Although he did not resolve the problem of solvability of the quintic by radicals, he did light upon a key idea, namely that the permutations of the roots of a polynomial equation are the "metaphysics" of its solvability by radicals. See Chapter 2.1.1 for details.

I try to give students a sense of Lagrange's ideas by showing how permutations of the roots of cubic and quartic equations help solve them by radicals [5], [6], [25]. Implicit in this is the notion of a group.

Although the Fundamental Theorem of Galois Theory is not needed to resolve the problem of solvability of the quintic, we do discuss the theorem, illustrating it with examples. It is a beautiful and important result, and it has nice applications—payoffs—aside from solvability by radicals.

Payoffs:
(a) Proofs of several important number-theoretic results: Fermat's "little" theorem, Euler's theorem, and Wilson's theorem. The proofs use only very elementary group theory [21].
(b) Classification of the regular polygons constructible with straightedge and compass. Although Galois theory yields a rather quick solution [23], the problem can be resolved using some field theory (cyclotomic extensions) and very elementary group theory [21].
(c) An essentially algebraic proof of the Fundamental Theorem of Algebra [23].
(d) Proof of the irrationality of expressions such as $\sqrt{3} + \sqrt{4} + \sqrt{72}$ [20].

Problem V: "Papa, can you multiply triples?"

This problem deals with extensions of the complex numbers to hypercomplex numbers, in particular the quaternions. The question in the title was asked by Hamilton's sons of their father to inquire whether he had succeeded, after years of effort, in obtaining an algebra of triples of reals analogous to the complex numbers. The problem bears on relations between arithmetic/classical algebra and abstract algebra, and it gives rise to the concepts of an algebra (not necessarily associative) and a division ring (a skew field). See Chapter 3.1 and Chapter 8.5.5.

To set the scene, I give the students a brief history of complex numbers. An important point to keep in mind here is that complex numbers arose in connection with the solution of the *cubic* rather than the quadratic [15].

Hamilton's quaternions—a noncommutative "number system"—was conceptually a most important development, for it liberated algebra from the canons of arithmetic (see Chapter 8.5.5). The history of their invention in 1843 is well documented and gives a rare glimpse of the creative process at work in mathematics [27].

Are there "numbers" beyond the quaternions? (What *is* a number, anyway?) Cayley's and, independently, Graves' octonions (8-tuples of reals) gave an affirmative answer, and raised the obvious question whether there are numbers beyond the octonions. A negative answer this time, given partly by Frobenius and C.S. Peirce, again independently [13]. Implicit in these ideas are the notions of division ring and algebra. See Chapter 3.1.

Payoffs:
(a) Ideas on quaternions can be used to prove Lagrange's four-squares theorem: Every positive integer is a sum of four squares [9], [10].
(b) Are complex numbers unavoidable in the solution of the so-called irreducible cubic? Yes. There is a proof using the considerable power of Galois theory [3], but the result can also be established by means of elementary field-extension theory [24].
(c) A relatively elementary proof of the Frobenius – C. S. Peirce theorem about the nonexistence of (associative) division rings of n-tuples of reals for $n \neq 1, 2, 4$ [13].

General remarks on the course

(a) The first and last problems, and probably also the second, are atypical in an abstract algebra course, but I have found them to be pedagogically enlightening and rich in algebraic ideas. Historically, they signalled the transition from classical to modern (abstract) algebra.
(b) The first problem begins with a "simple" numerical question. The idea is to ease students gently into the abstractions.
(c) While the sequence of topics in algebra books, and therefore in algebra courses, is usually from groups to rings to fields, our problems introduce students first to rings, then fields, and finally groups. I have found this order to be more effective. It leaves to the end the conceptually most difficult notion, that of a group, which students often find "unnatural."
(d) I have listed only five problems. It might be argued that this does not appear to be sufficient for an entire course. However, the problems are wide-ranging and rich in ideas, and are extendable in various directions, some of which are indicated in the various "payoff" sections.
(e) No textbook is used in the course. However, many references are given, both technical and historical, and students are expected to *read* some of them!
(f) The historical material used in the course comes mainly from secondary sources. Asking students (and instructors!) to read and assimilate primary sources would

make the course unreasonably difficult. The course is quite challenging as it is. And its objectives can be met using secondary sources.

(g) The course tries to deal with mathematical *ideas* in addition to the standard algebraic fare: the "why" and "what for" in addition to the "how." This is reflected in the assignments. Thus, aside from being asked to do the usual types of problems, for example, to show that the additive inverse in a ring is unique, students are expected to write "mini-essays" involving both historical and technical matters, for example, to discuss De Morgan's contribution to algebra and how it advanced abstract algebraic thinking.

To read independently in the mathematical literature, and to *write* about what they have read, are tasks which mathematics students are not—but should become— accustomed to.

References

1. W. W. Adams and L. J. Goldstein, *Introduction to Number Theory*, Prentice-Hall, 1976.
2. G. Birkhoff, Current trends in algebra, *Amer. Math. Monthly* 1973, **80**: 760–782.
3. G. Birkhoff and S. MacLane, *A Survey of Modern Algebra*, 5th ed., A K Peters, 1996.
4. D. M. Burton and D. H. Van Osdol, Toward the definition of an abstract ring, in *Learn from the Masters*, ed. by F. Swetz et al, Math. Assoc. of America, 1995, pp. 241–251.
5. A. Clark, *Elements of Abstract Algebra*, Dover, 1984.
6. R. A. Dean, *Elements of Abstract Algebra*, Wiley, 1966.
7. A. Fraenkel, Über die Teiler der Null und die Zerlegung von Ringen, *Jour. für die Reine und Angew. Math.* 1914, **145**: 139–176.
8. L. J. Goldstein, *Abstract Algebra: A First Course*, Prentice-Hall, 1973.
9. G. H. Hardy and E. M. Wright, *An Introduction to the Theory of Numbers*, Oxford Univ. Press, 1938.
10. I. N. Herstein, *Topics in Algebra*, Blaisdell, 1964.
11. D. Hilbert, Über den Zahlbegriff, *Jahresb. der Deut. Math. Verein.* 1900, **8**: 180–184.
12. A. Jones, S. A. Morris, and K. R. Pearson, *Abstract Algebra and Famous Impossibilities*, Springer-Verlag, 1991.
13. I. L. Kantor and A. S. Solodovnikov, *Hypercomplex Numbers*, Springer-Verlag, 1989. (Translated from the Russian by A. Shenitzer.)
14. I. Kleiner, The roots of commutative algebra in algebraic number theory, *Math. Magazine* 1995, **68**: 3–15.
15. I. Kleiner, Thinking the unthinkable: The story of complex numbers (with a moral), *Math. Teacher* 1988, **81**: 583–592.
16. M. Kline, *Mathematics in Western Culture*, Oxford Univ. Press, 1964.
17. N. H. McCoy and G. J. Janusz, *Introduction to Modern Algebra*, Wm. C. Brown, 1992.
18. H. Pollard and H. G. Diamond, *The Theory of Algebraic Numbers*, Math. Assoc. of America, 1975.
19. H. Pycior, George Peacock and the British origins of symbolical algebra, *Hist. Math.* 1981, **8**: 23–45.
20. I. Richards, An application of Galois theory to elementary arithmetic, *Advances in Math.* 1974, **13**: 268–273.
21. F. Richman, *Number Theory: An Introduction to Algebra*, Brooks/Cole, 1971.

22. I. Stewart, *Concepts of Modern Algebra*, Penguin, 1975.
23. I. Stewart, *Galois Theory*, 3rd ed., Chapman & Hall, 2004.
24. J. M. Thomas, *Theory of Equations*, McGraw Hill, 1938.
25. J. P. Tignol, *Galois Theory of Algebraic Equations*, Wiley, 1988.
26. B . L. van der Waerden, *A History of Algebra*, Springer-Verlag, 1985.
27. B . L. van der Waerden , Hamilton's discovery of quaternions, *Math. Magazine* 1976, **49**: 227–234.
28. P. L. Wantzel, Recherches sur les moyens de reconnaitre si un Problème de Géométrie peut se résoudre avec la règle et le compass, *Journ. de Math. Pures et Appl.* 1837, **2**: 366–372.

Biographies of Selected Mathematicians

8.1 Arthur Cayley (1821–1895)

Cayley was born at Richmond, in Surrey, England. He spent the first eight years of his life in St. Petersburg, Russia, where his father had settled as a merchant. When his father retired in 1829, the family moved to Blackheath, near London. Cayley went to a private school for six years, showing a great liking for and ability in numerical computations. At fourteen he entered King's College School in London. His father expected his son to pursue a career in business, but he showed a great aptitude for mathematics, and the Principal of the College persuaded his father to have Cayley study mathematics. He entered Trinity College, Cambridge, at the age of seventeen.

He performed brilliantly at Cambridge, graduating in 1842 as a Senior Wrangler (top honors) in the highly competitive Mathematical Tripos exams. He also won the greatly coveted Smith's prize, the first time it was awarded. He proceeded to an M.A. degree, and upon its completion in 1845 was awarded a Major Fellowship. This did not entail teaching, but it could be held only for a short time, unless one took holy orders. This Cayley was reluctant to do—not because of religious scruples but because he believed he was unsuitable for the post. Since no position as a mathematician was available to him in England, he left Cambridge in 1846 to study law, and was called to the Bar in 1849.

He spent the next fourteen years practicing law. His heart, however, was in mathematics, as a colleague noted:

> There is no doubt that had he remained at the Bar and devoted himself to its business, he could have made a great legal reputation and a substantial fortune: even as it was, some of his drafts have been made to serve as models. But the spirit of research possessed him; it was not merely will but an irresistible impulse that made the pursuit of mathematics, not the practice of law, his chief desire. To achieve this desire, he reserved with jealous care a due portion of his time; and he regarded his legal occupation mainly as the means of providing a livelihood [6].

Arthur Cayley (1821–1895)

The "due portion of his time" reserved for research sufficed for him to publish close to 300 mathematical papers during that period (1849–1863), including some of his best work. Earlier, in his student days, he had published over thirty articles. During his tenure as a lawyer he met Sylvester, who at that time worked as an actuary. They formed a life-long friendship, and both profited greatly from frequent discussions on mathematics, especially on the theory of invariants, which they are both credited with having founded (see below). Sylvester said of Cayley that he "habitually discourses pearls and rubies."

In 1863 Cambridge established a new Chair in mathematics—the Sadlerian—and offered it to Cayley. He promptly accepted, even though it meant a considerable reduction in income. That same year, at the age of forty two, he married. He had two children and a happy home life.

Cayley was caring, modest, kind, fair-minded, and unassuming. He was generous with his help and encouragement to young colleagues and was instrumental in furthering the higher education of women. Often consulted by the university administration on legal matters, he contributed willingly to the smooth running of the institution. He had a great capacity for work, but also took considerable interest in such "extra-curricular" activities as painting, literature, architecture, travel, and mountaineering. He had a knowledge of French, German, Italian, Latin, and Greek, and published in leading French and German journals.

Cayley's mathematical contributions ranged over all then-existing areas of pure mathematics, including theoretical dynamics and mathematical astronomy. He was among the three most prolific mathematicians of all time, the other two being Euler and Cauchy. His Collected Papers comprise thirteen volumes, containing about 1000 articles. He wrote one book, *Treatise on Elliptic Functions*, which helped familiarize the English-speaking world with this fundamental subject.

Cayley's main contributions were in various areas of algebra (invariants, groups, matrices) and geometry (especially projective geometry). We will stress the algebra.

8.1.1 Invariants

The notion of invariance is of course fundamental in mathematics (as it is in science). Gauss was among the first to explicitly recognize invariance in his number-theoretic investigations of binary quadratic forms, $f(x, y) = ax^2 + bxy + cy^2$. He showed that the discriminant $D = b^2 - 4ac$ of the form f is invariant under a linear transformation of its variables. More generally, if f is transformed by a linear mapping T of the variables x, y into the form $g(u, v) = ku^2 + luv + mv^2$, then any function I of the coefficients of f which satisfies the relation $I(k, l, m) = t^k I(a, b, c)$ is called an *invariant* of f under T (t denotes the determinant of the matrix of T). A corresponding definition applies to a form (i.e., a homogeneous polynomial) of any degree in any number of variables.

Invariants also proved important in geometry, especially projective and algebraic geometry, which sought properties of figures invariant under projective and birational transformations, respectively. These were the intrinsic properties of the respective geometries.

Cayley was inspired to study invariants by an 1841 paper of Boole on the subject, entitled "Exposition of a general theory of linear transformations" (see [11]). Initially, his focus was on geometry. (Since a form f of degree m in n variables is a homogeneous polynomial, $f = 0$ represents a curve, a surface, ...) Soon, however, he adopted an abstract point of view and, when he was twenty-two, formulated the first of two basic problems of invariant theory:

> [To] find all the derivatives [invariants] of any number of functions [forms], which have the property of preserving their form unaltered after any linear transformation of the variables [3]. [The term "invariant" was coined by Sylvester.]

In 1846 he devised a method for generating invariants of binary forms. For example, he showed that $g = ae - 4bd + 3c^2$ is an invariant of the binary quartic form $ax^4 + 4bx^3y + 6cx^2y^2 + 4dxy^3 + ey^4$ under unimodular transformations (linear transformations of determinant 1), where a, b, c, d are complex numbers. Over a twenty-five-year period, starting in 1854, Cayley wrote ten papers, all entitled "Memoir upon quantics [forms]" (numbered consecutively), computing invariants of various forms.

The second major problem of the theory was to find, if possible, a finite number of invariants of a given form which would generate *all* invariants of the form, that is, such that every invariant is a combination of these invariants. Such a finite set was called a "complete system" (later also a "basis") of invariants for the form. For example, for the above binary quartic, Cayley showed that g (as above) and $h =$ the determinant of $(a, b, c), (b, c, d), (c, d, e)$ (the triples are the rows of the determinant) are a complete system of invariants under unimodular transformations. One of his important results was finding a basis for all *binary* forms of degree six or less. (Finding covariants of forms was a related problem, but we will not deal with it.)

Between the 1840s and 1880s invariant theory became a major branch of algebra, with connections to geometry, number theory, and linear algebra. To some it constituted the modern algebra of the period. Sylvester claimed that "all algebraic inquiries, sooner or later, end at the Capitol of modern algebra over whose shining portal is inscribed the Theory of Invariants." The subject attracted, among others, Jordan and Hermite in France, and Clebsch, Gordan, and Hesse in Germany.

A great many invariants were found for specific forms. The pressing problem became to find a basis for various forms. In 1868 Gordan proved that every *binary* form, of *any degree*, has a *finite* basis. (Cayley had earlier conjectured that binary forms of degree greater than six had an infinite basis.) His proof of this important result was computational and difficult. In 1888 Hilbert rephrased the problem in terms of the newly emerging concepts of rings and ideals, and showed that a polynomial ring in finitely many variables has a finite basis. This is the so-called Hilbert Basis Theorem. It implies that every form—of any degree, in any number of variables—has a finite basis. The result seemed to have "killed" invariant theory [5]. But it reemerged, with vigor, in the second half of the twentieth century [12].

8.1.2 Groups

In 1854 Cayley gave the first abstract definition of a group (see Chapter 2.3; for a critical look at Cayley's definition see [9]). His motivation derived first from the work on permutation groups by Cauchy, and especially by Galois, and second from the contributions to algebra of British mathematicians, among them Peacock, de Morgan, Hamilton, and Boole. Wussing [14] claims that Cayley was also influenced in his work on groups by invariant theory.

The immediate ancestor of the abstract notion of a group was Galois, as Cayley acknowledged: "the idea of a group as applied to permutations or substitutions is due to Galois, and the introduction of it may be considered as marking an epoch in the progress of the theory of algebraical equations." As for British mathematicians, in the period 1830s–1850s they founded "symbolical algebra," which asserted—initially only in theory—that what mattered in algebra is not the meaning of the symbols involved in algebraic expressions but their laws of combination (see Chapter 1.8). Later they introduced various systems with properties which differed in various ways from those of the traditional number systems. These included quaternions and biquaternions (Hamilton), triple algebras (de Morgan), octonions (Graves and Cayley), Boolean algebras (Boole), and matrices (Cayley). See Chapter 3.1.

Cayley was a pure mathematician who sought generality and (presumably) wanted to unify some of these and other examples under one roof. In fact, as instances of groups he listed permutations, quaternions (under addition), invertible matrices (under multiplication), binary quadratic forms (under composition of forms as defined by Gauss in 1801—see Chapter 3.2), groups which arose in the theory of elliptic functions, and two groups, of orders eighteen and twenty-seven, respectively, defined by generators and relations.

Cayley's 1854 definition of a group attracted little attention. The only major source of group theory at the time was algebraic equations, so there was little need to generalize. Also, abstraction and axiomatics were not in vogue in the mid-nineteenth century. However, his work in this area was the first example of a major algebraic system to be axiomatized. Moreover, it was among the first examples of what Eves called "formal axiomatics" [4], in which an axiom system abstracts *distinct* mathematical objects. This is in contrast to "material axiomatics," which seeks to describe by means of axioms an essentially unique mathematical system (e.g., the euclidean plane, the real numbers).

In 1878 Cayley returned to the subject, writing four short papers on abstract groups. The mathematical climate was entirely different at this time. Group theory was now related, in addition to the theory of equations, to geometry, number theory, and analysis, and the abstract point of view had penetrated other areas of algebra (cf. B. Peirce's work on noncommutative algebra and Dedekind's on ideal theory; see Chapter 3.1 and 3.2). Cayley's 1878 work inspired several mathematicians working in group theory, especially Hölder, von Dyck, and Burnside.

8.1.3 Matrices

"One could say that the subject of matrices was well developed before it was created," notes Kline [8]. Indeed, many of the properties of matrices were apparent from the earlier study of systems of linear equations and determinants. As Cayley put it in the introduction to his 1858 paper on matrices, the more substantial of two articles he wrote on the topic:

> A set of quantities arranged in the form of a square ... is said to be a matrix. The notion of such a matrix arises naturally from an abbreviated notation for a set of linear equations, ... and the consideration of such a system of equations leads to most of the fundamental notions in the theory of matrices.

He continued:

> It will be seen that matrices (attending only to those of the same order) comport themselves as single quantities; they may be added, multiplied or compounded together, etc: the law of the addition of matrices is precisely similar to that for the addition of algebraical quantities; as regards their multiplication (or composition), there is the peculiarity that matrices are not in general convertible [commutative]; it is nevertheless possible to form powers (positive or negative, integral or fractional) of a matrix, and hence to arrive at the notion of a rational and integral function, or generally any algebraical function, of a matrix. I obtain the remarkable theorem that any matrix whatever satisfies an algebraical equation of its own order, the coefficient of the highest power being unity, and those of the other powers functions of the terms of the matrix, the last coefficient being in fact the determinant [1].

The "remarkable theorem" is of course the Cayley–Hamilton theorem, which he proved only for 2×2 matrices. "The general theorem," he said, "...will be best understood by a complete development of a particular case." He noted further that he had "verified the theorem in the next simplest case of a matrix of the order 3," and that he had "not thought it necessary to undertake the labor of a formal proof of the theorem in the general case of a matrix of any degree."

Here are some of the other concepts he introduced in this paper: the zero and identity matrices, inverses, symmetric and skew symmetric matrices, transpose of a matrix, and rectangular matrices. Among the results he proved, all essentially elementary, are the following: every matrix is a sum of a symmetric and skew-symmetric matrix, the transpose of a product of matrices is the product of their transposes, in reverse order. He also showed how to find the inverse of a matrix by the cofactor method, and determined all (2×2) matrices commuting with a given matrix. Finally, he related matrices to quaternions, as follows: if L and M are 2×2 matrices such that $LM = -ML$ and $L^2 = M^2 = -1$, then letting $N = ML$, we get $N^2 = -1, MN = -NM, NL = -LN$, "which is a system of relations precisely similar [we would say isomorphic] to that in the theory of quaternions" [1].

It is noteworthy that all of Cayley's results in this paper are verified ("proved") only for 2×2 or 3×3 matrices. This is how he viewed the matter: "For conciseness, the matrices written down at full length will in general be of the order 3, but it is to be understood that the definitions, reasonings, and conclusions apply to matrices of any degree whatever" [1]. This attitude was in keeping with his overall approach to proofs: as long as he had convinced himself of the validity of a result, he saw no need for a formal proof.

Cayley's work on matrices had little early impact, perhaps because he did not relate it to significant mathematics or to applications. The importance of his contribution in this area was the consideration of matrices as single entities and the development of an algebra of matrices.

8.1.4 Geometry

The nineteenth century saw an explosive growth in geometry. In 1822 Poncelet rediscovered projective geometry. (Desargues in the seventeenth century was the discoverer, but his work went unnoticed.) Over the course of the next three decades much work was done on the subject, using both synthetic and analytic methods. In parallel, noneuclidean geometry was introduced and explored. Attempts began to put some order in the various geometries.

Poncelet developed projective geometry using notions from euclidean geometry (length and angle). To him projective geometry was a subset of euclidean geometry. Others began to believe that projective geometry is more basic that euclidean geometry. The sense emerged that euclidean geometry is, in fact, a subgeometry of projective geometry. This is precisely what Cayley proved in 1859, in his sixth "Memoir on quantics."

He accomplished his goal by defining length and angle in terms of what he called "absolutes." In two dimensions these are conics, and in three dimensions quadric surfaces. The formulas for length (the so-called "Cayley metric") and angle were given analytically, in terms of bilinear and quadratic forms, and were thereby defined only in terms of projective concepts. By picking a specific conic as absolute, he obtained the euclidean formulas for distance and angle. This showed that (as he put it) "metrical geometry is part of projective geometry." See [8], [13] for details.

However, Cayley failed to exploit his metric to the fullest, in that he did not relate it to noneuclidean geometry. Perhaps this was because of his ambivalent attitude toward that geometry. The gap was filled by Klein in 1871. By scrutinizing the Cayley metrics associated with various absolutes he discovered plane and solid hyperbolic and elliptic geometries, which are also subgeometries of projective geometry. This work also yielded a model of the hyperbolic plane, and led Klein to the formulation of the Erlanger Program. See [14].

To conclude our discussion of Cayley's contribution to geometry, we mention three of his numerous results on what can perhaps be described as classical geometry. (His geometric studies were often intimately connected with his work on invariants.) He focused on curves and surfaces, but was not averse, as early as 1845 (when he was twenty-four), to speak of *n-dimensional geometry*. In particular, he found geometric analogy useful in his algebraic work. In 1854, in the first of his ten memoirs on quantics, he asserted: "I consider that there is an ideal space of any number of dimensions, but of course, in the ordinary acceptation of the word, space is of three dimensions" [14]. Clerk Maxwell, in an excerpt from a poem he wrote in Cayley's honor, put it thus: "His soul too large for vulgar space, in n dimensions flourished unrestricted." Now to the three results:

(a) In a paper entitled "On the triple tangent planes of surfaces of the third order" Cayley showed that there are 27 lines on a cubic surface. "The discovery opened up the theory of surfaces, and attracted much international attention" [7].

(b) If a plane curve of degree r passes through two curves of degrees m and n, respectively, where $r > m, n$, the number of conditions needed to determine the curve is $mn - (m + n - r - 1)(m + n - r - 2)$. This result, known as the Cayley–Bacharach intersection theorem, was later generalized by Max Noether.

(c) In the later seventeenth century Newton gave a classification of cubic curves, and in 1835 Plücker gave another. In a paper "On the classification of cubic curves" Cayley "expound[ed] the principles of the two classifications, and brought them into comparison with one another, and entering into the discussion with full minuteness, he obtain[ed] the exact relation of the two classifications to one another—a result of great value in the theory" [6].

8.1.5 Conclusion

We come to the end of our account of Cayley. The following quotation from A. R. Forsyth, Cayley's successor in the Sadlerian chair at Cambridge, gives expression to what is a major element in his work:

> As is often (and naturally) the case with the discoverer of a fertile subject, Cayley himself did not explain or foresee the full range of applications of his new ideas [6].

Indeed, in none of the topics we have discussed did Cayley exploit the full potential of an idea that he had introduced (which can probably be said of most if not all mathematicians). But his ideas often inspired others, who brought them to fruition.

In the theory of invariants, in which Cayley published extensively, he proved only a very special case of the fundamental basis theorem. The major result was proved by Gordan, and was later extended by Hilbert using ideal-theoretic language. But as Wussing claims, the problem of seeking a basis for invariants "strongly influenced Cayley's formulation of the group concept and, on the other hand, was one of the historical roots of ideal theory" [14]. Invariant theory, which we recall was initiated by Cayley and Sylvester, also "appears in retrospect to have been a transitional stage on the way to the later, explicitly group-theoretic classification of the whole edifice of geometry" [14].

In his work on groups and matrices (recall that he had introduced both concepts), Cayley only scratched the surface. For one reason or another, he did not pursue their extensive and deep ramifications. In geometry he proved (among many things) the fundamental result that euclidean geometry is a subgeometry of projective geometry, but missed the case of noneuclidean geometry.

It is not our intention to leave the reader with a negative impression of Cayley's accomplishments. He was undoubtedly a ranking mathematician of the nineteenth century, "the first English mathematician to achieve international recognition since Newton, and the first to open up Continental mathematics to an Anglophone audience" [7]. Many honors came his way. He was a fellow of the Royal Society of Edinburgh, of the Royal Irish Academy, and of the Royal Astronomical Society. He was also either a Fellow or a foreign corresponding member of most of the scientific societies of the Continent, including the French Institute, and the Academies of Berlin, Göttingen, St. Petersburg, Milan, Leyden, Upsala, and Hungary. He was President of the Cambridge Philosophical Society, of the Royal Astronomical Society, and of the British Society for the Advancement of Science. He received from the Royal Society the Royal Medal and the Copley Medal, the highest scientific distinction that was in its power to confer.

Finally, here is a brief tribute from Professor Forsyth, who wrote an extensive obituary of Cayley:

> With a singleness of aim . . . he persevered to the last in his nobly lived ideal. His life had a significant influence on those who knew him: they admired his character as much as they respected his genius; and they felt that, at his death, a great man had passed from the world [6].

References

1. A. Cayley, A memoir on the theory of matrices, in Cayley's *Collected Mathematical Papers*, Cambridge Univ. Press, 1895, vol. II, pp. 475–496.

2. T. Crilly, *Arthur Cayley: Mathematician Laureate of the Victorian Age*, Johns Hopkins University Press, 2006.

3. T. Crilly, Invariant theory, in *Companion Encyclopedia of the History and Philosophy of the Mathematical Sciences*, ed. by I. Grattan-Guinness, Routlage, 1994, vol. 1, pp. 787–793.

4. H. Eves, *Great Moments in Mathematics (After 1650)*, Math. Assoc. of Amer., 1981.

5. S. Fisher, The death of a mathematical theory: a study in the sociology of knowledge, *Arch. Hist. Exact Sci.* 1966, **3**: 137–159.

6. R. Forsyth, Obituary of Cayley, in Cayley's *Collected Mathematical Papers*, Cambridge Univ. Press, 1895, vol. VIII, pp. vii–xlvi.

7. J. Gray, Arthur Cayley (1821–1895), *Math. Intelligencer* 1995, **17**(4): 62–63.

8. M. Kline, *Mathematical Thought from Ancient to Modern Times*, Oxford Univ. Press, 1972.

9. P. M. Neumann, What groups were: a study of the development of the axiomatics of group theory, *Bull. Austral. Math. Soc.* 1999, **60**: 285–301.

10. D. North, Cayley, Arthur, in *Dictionary of Scientific Biography*, ed. by C. C. Gillispie, Charles Scribner's Sons, 1981, vol. 3, pp. 162–170.

11. K. H. Parshall, Toward a history of nineteenth-century invariant theory, in *The History of Modern Mathematics*, ed. by D. E. Rowe and J. McCleary, Academic Press, 1989, vol. 1, pp. 157–206.

12. G.-C. Rota, Two turning points in invariant theory, *Math Intelligencer* 1999, **21**(1): 20–27.

13. A. Shenitzer, The Cinderella career of projective geometry, *Math. Intelligencer* 1991, **13**(2): 50–55.

14. H. Wussing, *The Genesis of the Abstract Group Concept*, MIT Press, 1984. (Translated from the German by A. Shenitzer.)

8.2 Richard Dedekind (1831–1916)

The nineteenth century was a golden age in mathematics. Entirely new subjects emerged (e.g., abstract algebra, noneuclidean geometry, set theory, complex analysis) and old ones were radically transformed (e.g., real analysis, number theory, geometry). Just as important, the spirit of mathematics, the way of thinking about it and doing it, changed fundamentally, even if gradually.

Mathematicians turned more and more for the genesis of their ideas from the sensory and empirical to the intellectual and abstract. Witness the introduction of noncommutative algebras, noneuclidean geometries, continuous nowhere differentiable functions, space-filling curves, n-dimensional spaces, and completed infinities of different sizes. Cantor's dictum that "the essence of mathematics lies in its freedom" became a reality, though one to which many mathematicians took strong exception.

Other pivotal changes were the emphasis on *rigorous* proof and the acceptance of nonconstructive existence proofs, the focus on concepts rather than on formulas and algorithms, the stress on generality and abstraction, the resurrection of the axiomatic method, and the use of set-theoretic modes of thinking. Dedekind was an exemplary practitioner of many of these new undertakings; in fact, he initiated several of them—as we shall see.

Dedekind was born in Brunswick, Germany (also the birth place of Gauss). His father was a lawyer and a professor at the Collegium Carolinum (an educational institution between a high school and a university), and his mother the daughter of a professor at the same college. The youngest of four children, he never married, living for many years with his sister until her death in 1914.

Between the ages of seven and sixteen Dedekind attended the local gymnasium, studying physics and chemistry. However, he found these subjects unsatisfactory since they lacked logical structure! In 1848, at sixteen, he entered the Collegium Carolinum (which Gauss had earlier attended). There he mastered the elements of analytic geometry, calculus, algebra, and mechanics. He was thus well prepared when he entered the University of Göttingen two years later.

Richard Dedekind (1831–1916)

The only mathematician of note at Göttingen in 1850 was Gauss. Dedekind took courses on the elements of number theory, on differential and integral calculus, on hydraulics, and on experimental physics. He found Gauss' lectures on the method of least squares and on advanced geodesy inspiring, but on the whole he felt that these studies did not prepare him well for mathematical research. So, following his PhD, he undertook intensive study on his own for two years to gain deep knowledge of such current subjects as elliptic functions, recent developments in geometry, and advanced algebra and number theory.

He got his doctorate under Gauss in 1852 (at the age of twenty-one) on the topic of Eulerian integrals. Gauss noted about the dissertation that "the author evinces not only a very good knowledge of the relevant field, but also such an independence as augurs favorably for his future achievement" [2].

The *Habilitationsschrift* is a probationary (research) lecture traditionally given by academics in German universities before they are entitled to teach. Dedekind gave his probationary lecture in 1854. It dealt with extensions of the number system, beginning with the natural numbers, each system generated from the one before it in a systematic way; for example, the negative integers from the natural numbers, and the rationals from the integers. These were entirely new issues which had not arisen before Dedekind's time. He expressed, in particular, his dissatisfaction with the contemporary presentation of the irrational numbers (see the section below on real numbers). The elaboration of the ideas in this lecture, with the focus on the primacy of numbers in mathematics, formed a vital part of his mathematical research program over the next thirty years.

The first courses Dedekind taught at Göttingen were on probability and geometry. In 1856 he gave courses on Galois theory and group theory, probably the first university teacher to lecture on these important new subjects. During these two years (1854–56) he also *attended* lectures, on abelian and elliptic functions by Riemann, who had come to Göttingen in 1851 to pursue doctoral studies with Gauss, and on number theory, potential theory, and analysis by Dirichlet, who came to Göttingen in 1855 to succeed Gauss upon his death. Dedekind formed lasting friendships with both Riemann and Dirichlet and was influenced both by their mathematics and by their approach to the subject, which focused on getting at the underlying concepts of a theory rather than the computations. Dirichlet in particular made a "new man" out of him, he said.

In 1858 Dedekind was appointed professor at the prestigious Zürich Polytechnic (now the ETH). He was recommended for this position by Dirichlet, who, in addition to praising his mathematical abilities, called him "an exceptional pedagogue." He stayed at the Polytechnic four years, and in 1862 became professor at the Brunswick Polytechnic in his home town, where he spent the last fifty years of his life.

Among Dedekind's contributions to mathematics three stand out: his founding of algebraic number theory (1871), his definition of the real numbers in terms of what are now known as Dedekind cuts (1872), and his definition of the natural numbers in terms of sets (1888). We discuss each in turn. (His work with Weber (1882) on algebraic function fields is also noteworthy.) We should note that although we give here the formal publication dates of the respective works, Dedekind had thought about the basic outline of these works since the 1850s. But he was a perfectionist and would not publish until he was satisfied that he got at the fundamental ideas underlying the theories. It is also noteworthy that the three contributions have a broad common theme: "numbers"—algebraic numbers, real numbers, and natural numbers, respectively. Indeed, Dedekind believed in the primacy of numbers in mathematics, in particular that algebra and analysis should be based on the natural numbers. The following reflections on the supremacy of number come from his *Nachlass*:

> Of all the aids which the human mind has yet created to simplify its life—that is, to simplify the work in which thinking consists—none is so momentous and inseparably bound up with the mind's most inward nature as the concept of *number*. Arithmetic, whose sole object is this concept, is already a science

of immeasurable breadth, and there can be no doubt that there are absolutely no limits to its further development; and the domain of its application is equally immeasurable, for every thinking man, even if he does not clearly realize it, is a man of numbers, an arithmetician [12].

8.2.1 Algebraic Numbers

The study of number theory goes back several millennia. In the eighteenth century the subject had two brilliant exponents in Euler and Lagrange, but its systematic investigation, which made it into a major branch of mathematics, began with Gauss' *Disquisitiones Arithmeticae*. Several of the foremost mathematicians of the nineteenth century, including Dirichlet, Riemann, Kummer, Kronecker, Hermite, Hilbert, and Dedekind, made major contributions to the field. Gauss aside, Dedekind's was arguably the most significant, in several ways: its formulation, its grand conception, its fundamental new ideas, its modern spirit, and its impact.

Algebraic number theory arose mainly from the investigation of three fundamental problems in classical number theory: Fermat's Last Theorem, reciprocity laws, and binary quadratic forms. Although these had their roots in the seventeenth and eighteenth centuries, they began to be intensively investigated only in the nineteenth. The strategy that started to emerge was to embed the domain of integers, in terms of which these problems were formulated, in domains of what came to be known as algebraic integers. Two basic questions then came to the fore: what exactly *are* such domains?, and can one formulate a unique factorization theorem for them analogous to that for the integers? See Chapter 3.2 for details.

Pioneering work on the topic was done in the 1840s by Kummer, who showed that in the domain of cyclotomic integers every element is a unique product of ideal primes. It is mainly this work that inspired Dedekind. His two main tasks were to generalize the domain of cyclotomic integers to much wider domains (needed in number-theoretic problems), and to extend Kummer's mysterious ideal numbers to these domains. Dedekind's initial thoughts on these issues needed twenty years to mature. He achieved his aim in 1871, in the now famous Supplement X to the 2nd edition of Dirichlet's *Vorlesungen über Zahlenteorie*.

The *Vorlesungen* went through four editions, with Dedekind appending improved Supplements to the 3rd and 4th, of 1879 and 1894 respectively. In these Supplements Dedekind created a beautiful new edifice—algebraic number theory—in which he introduced such fundamental algebraic concepts as ring, field, ideal, and module. In 1877 he wrote a French version of ideal theory in order to present his ideas to a general mathematical audience (now available in an English translation, with commentary, by Stillwell [6]).

In Chapter 3.2 we have given details of Dedekind's creation of algebraic number theory. Here we mention only his definitions of field and ideal—both crucial in that theory—in order to focus on some of his *methodological* achievements.

A system [set] F of real or complex numbers is called a *field* if the sum difference, product, and quotient of any two numbers of F belong to F.

A subset I of the integers R of an algebraic number field K is an *ideal* of R if it has the following two properties:

(i) If $a, b \in I$, then $a \pm b \in I$.

(ii) If $a \in I, c \in R$, then $ac \in I$.

Dedekind introduced here two fundamental innovations:

(i) Use of the axiomatic method *in algebra*, although in the concrete setting of the complex numbers, which was all that was needed for algebraic number theory. This influenced Hilbert and especially Noether, and became a staple of twentieth-century mathematics.

(ii) Use of set-theoretic language and of the completed infinite. (Observe that Dedekind's fields and ideals are infinite sets.) This predated Cantor's work on sets later in the 1870s.

We want to focus for a moment on Dedekind's abstract definition of an ideal. Earlier in his thinking he encountered ideals as sets given by a finite number of generators, but this notion depended on a particular representation of the ideal, which Dedekind did not find satisfactory. He was looking for *intrinsic* properties of ideals. On the other hand, Kronecker viewed ideals *precisely* as linear combinations of generators. To him Dedekind's axiomatic definition was anathema.

As we mentioned, the conceptual focus adopted by Dedekind was promoted earlier by his two colleagues and friends, Dirichlet and Riemann. Speaking of Dirichlet's work, and noting the famous "Dirichlet Principle" in analysis, Minkowski referred to "the other principle of Dirichlet" as the view that mathematical problems should be solved through a minimum of blind calculation and a maximum of forethought. Edwards claims that Dedekind introduced the notion of an ideal abstractly "in an effort to do for number theory what Riemann had done for function theory—which Dedekind saw as the elimination of formulas and calculations in favor of intrinsic properties" [10].

Indeed, one of the consequences of the introduction of ideals was a conceptual understanding of Gauss' theory of binary quadratic forms, expressions of the form $f(x, y) = ax^2 + bxy + cy^2$, where a, b, c are integers. The main problem of the theory was: given a form f, find all integers m that can be represented by f, that is, for which $f(x, y) = m$. In the *Disquisitiones* Gauss developed a comprehensive and beautiful theory of such forms. Most important was his definition of the *composition* of two forms. This was subtle and very difficult to describe. Attempts to gain conceptual insight into Gauss' theory of composition of forms inspired the efforts of some of the best mathematicians of the time, including Dirichlet and Kummer. Dedekind succeeded by associating with each binary quadratic form an ideal, and showing that composition of forms corresponds to multiplication of ideals. As he put it in a related context:

It is preferable, as in the modern theory of functions, to seek proofs based immediately on fundamental characteristics rather than on calculation, and indeed to construct the theory in such a way that it is able to predict the results of calculation [6].

It is noteworthy that the *product of ideals*, which Dedekind defined, is an operation on *sets* rather than on elements. This was a novel and important idea, which Dedekind was to use again in his definition of the real numbers (see below).

In conclusion, Dedekind's work in the Supplements to Dirichlet's *Vorlesungen über Zahlentheorie* founded a new subject—algebraic number theory. It also embodied a breakthrough in the evolution of abstract algebra. His approach and methods were revolutionary. Bourbaki called the work "magisterial," Landau said it "brought order to chaos, and light to the deepest darkness," and Noether noted that "its style of thought now [the 1920s] permeates the entirety of modern algebra." Despite the high praise from such distinguished quarters, Dedekind's ideal theory did not get a positive reception until the 1890s. Most nineteenth-century mathematicians were not prepared for its modern spirit.

8.2.2 Real Numbers

The real numbers were viewed throughout history variously as magnitudes, ratios of magnitudes, quantities, infinite decimals, or points on the line. None of these definitions was rigorous, nor were the properties of the real numbers explicitly formulated. This became a pressing issue for Dedekind already in 1858, when he began to teach calculus upon his appointment to the Zürich Polytechnic. The following rather long quotation reveals the prevailing state of affairs and Dedekind's thinking on the matter:

> My attention was first directed toward the considerations which form the subject matter of this pamphlet in the autumn of 1858. As professor in the Polytechnic school in Zürich I found myself for the first time obliged to lecture upon the elements of the differential calculus and felt more keenly than ever before the lack of a really scientific formulation for arithmetic. In discussing the notion of the approach of a variable magnitude to a fixed limiting value, and especially in proving the theorem that every magnitude which grows continually, but not beyond all limits, must certainly approach a limiting value, I had recourse to geometric evidence. Even now such resort to geometric intuition in a first presentation of the differential calculus I regard as exceedingly useful from the didactic standpoint, and indeed indispensable if one does not wish to lose too much time. But that this form of introduction into the differential calculus can make no claim to being scientific, no one will deny. For myself this feeling of dissatisfaction was so overpowering that I made the fixed resolve to keep meditating on the question till I should find a purely arithmetic and perfectly rigorous foundation for the principles of infinitesimal analysis. The statement is so frequently made that the differential calculus deals with continuous magnitude, and yet an explanation of this continuity is nowhere given; even the most rigorous expositions of the differential calculus do not base their proofs upon continuity but, with more or less consciousness of the fact, they either appeal to geometric notions or those suggested by geometry, or depend upon theorems which are never established in a purely arithmetic manner. Among these, for example, belongs

the above-mentioned theorem, and a more careful investigation convinced me that this theorem, or any one equivalent to it, can be regarded in some way as a sufficient basis for infinitesimal analysis. It then only remained to discover its true origin in the elements of arithmetic and thus at the same time to secure a real definition of the essence of continuity. I succeeded on Nov. 24, 1858 [8].

To provide some context, Cauchy gave a rigorous presentation of the calculus based on the concept of limit in a seminal work begun in 1821. But he left unresolved a number of foundational issues. Since the real numbers are in the foreground or background of much of analysis, and were viewed *geometrically* by Cauchy and his contemporaries, they resorted to intuitive geometric arguments to establish a number of the fundamental results of analysis, for example, the Intermediate Value Theorem. Dedekind found this unacceptable. (So did several other mathematicians around 1872, in particular Cantor, Weierstrass, and Heine. Each gave a rigorous but different presentation of the real numbers.)

The details of Dedekind's definition of the reals in terms of "cuts," outlined in his pamphlet of 1872, *Continuity and Irrational Numbers*, are well known. Briefly, a Dedekind cut is a partition (A, B) of the rationals into two sets A and B such that every element of A is less than each element of B. Each rational number r gives rise to a cut (A, B) in a natural way $(A = \{x \in Q : x \leq r\}, B = \{x \in Q : x > r\})$, but there are (infinitely many) cuts not given by any rational number (e.g., $A = \{x \in Q : x^2 \leq 2\}$, $B = \{x \in Q : x^2 > 2\}$). The real numbers are defined to be the totality of all cuts. (Dedekind said that the real numbers "correspond" to the cuts, while we say they *are* the cuts.)

Dedekind showed that the newly created numbers possess "continuity" (*we* call it "completeness"), namely that if we repeat the process of forming cuts by partitioning the reals into two classes (as above) nothing new results. He then defined order and addition for the reals, and noted that:

> Just as addition is defined, so can the other operations of the so-called elementary arithmetic be defined, viz., the formation of differences, products, quotients, powers, roots, logarithms, and in this way we arrive at real proofs of theorems (as, e.g., $\sqrt{2}\sqrt{3} = \sqrt{6}$), which to the best of my knowledge have never been established before [8].

We have noted that Dedekind used infinite sets in his 1871 work on ideal theory. A year later he used them to define the real numbers: a real number is a pair of infinite sets of rationals. Moreover, "the elusive concept of a real number was, in Dedekind's view, made *concrete* by being viewed as a *cut* of the set of rationals" [10]. The entire development in his booklet *Continuity and Irrational Numbers* is in the language of sets. For example, addition of real numbers is defined as addition of cuts. Here, as in *Supplement X* on algebraic numbers, Dedekind defined an operation on *sets*. The language of sets will take center-stage in his work on the natural numbers (see below).

Dedekind devoted the final section of his work on real numbers "to explain the connection between the preceding investigations and certain fundamental theorems of infinitesimal analysis" [8]. In particular, he proved "one of the most important theorems [of the subject]" [8], that an increasing, bounded sequence of real numbers has a limit. He noted that the result is equivalent to the completeness of the real numbers. This work, in conjunction with Weierstrass' on the $\varepsilon - \delta$ definition of the limit, gave a "purely arithmetic and perfectly rigorous foundation for the principles of infinitesimal analysis" (see the long quotation above), thus completing what has been called "the arithmetization of analysis" [4].

8.2.3 Natural Numbers

Recall that already in his *Habiltationsschrift* of 1854 Dedekind laid out a general plan to construct the various number systems starting from the natural numbers and terminating in the real numbers. He published his definition of the natural numbers in a pamphlet of 1882, *Was sind und was sollen die Zahlen* (*What are Numbers and What are they Good For?*, mistranslated in [8] as *The Nature and Meaning of Numbers*). He noted that "the design of such a presentation I had formed before the publication [in 1872] of my paper on *Continuity* [*and Irrational numbers*]" [8].

The monograph did not win the universal praise of contemporary mathematicians. Even Dedekind anticipated potential misgivings about the unorthodox, abstract nature of the presentation:

> This memoir can be understood by any one who possesses what is usually called good common sense; no technical philosophic, or mathematical, knowledge is in the least required. But I feel conscious that many a reader will scarcely recognize in the shadowy forms which I bring before him his numbers, which all his life long have accompanied him as faithful and familiar friends; he will be frightened by the long series of simple inferences corresponding to our step-by-step understanding; by the matter-of-fact discussion of the chains of reasoning on which the laws of numbers depend, and will become impatient at being compelled to follow our proofs for truths which to his supposed inner consciousness seem at once evident and certain [e.g., the commutative law of addition]. On the contrary, in just this possibility of reducing such truths to others more simple, no matter how long and apparently artificial the series of inferences, I recognize a convincing proof that their possession or belief in them is never given by inner consciousness but is always gained only by a more or less complete repetition of the individual inferences [8].

By reducing "truths to others more simple" Dedekind meant reducing the properties of integers to those of sets and mappings (a presentiment of the logicist school). Thus the first twenty or so pages of the tract deal exclusively with the latter subjects. The first sentence reads: "In what follows I understand by a *thing* [an element] every object of our thought" [8]. Dedekind went on to define sets: "It very frequently happens that different things, $a, b, c, \ldots$ for some reason can be considered from a common point of view, can be associated in the mind, and we say that they form a *system* [set] S; we call the things $a, b, c, \ldots$ *elements* of the system S" [8]. He also defined equality of sets, subsets, unions, intersections, mappings of sets, and composition of maps—all in the modern spirit in which they are presented today.

He then introduced the notion of a "chain," in terms of which he defined the natural numbers. (Given a mapping φ of a set K into itself, K is called a *chain* if $\varphi(K)$ is a proper subset of K.) In a letter in 1890 to a school principal he gave a very clear summary of the main ideas in his monograph. There he defined the natural numbers N avoiding the notion of chain: Given any set S, a one-one mapping φ of S, and a distinguished element, call it 1, not in $\varphi(S)$, N is the intersection of all subsets K of S such that (i) $1 \in K$, (ii) $n \in K$ implies $\varphi(n) \in K$. The principle of mathematical induction followed as a *theorem* [7].

Among other theorems, he proved the following: (a) There exist infinite sets (!). (He defined an infinite set as one which has a proper subset of the same cardinality as the set itself.) (b) Every infinite set contains a copy of the natural numbers. (c) The set of natural numbers is unique, up to isomorphism.

Dedekind's memoir *Was sind und was sollen die Zahlen* was the most explicit of his works in its use of set-theoretic notions, which became central in twentieth-century mathematics. It inspired Peano in his axiomatic definition (in 1889) of the natural numbers and Zermelo in his search (in the 1900s) for an axiom system for sets. See [12], [13].

8.2.4 Other Works

We mention briefly several of Dedekind's other contributions.

(a) Algebraic geometry
Dedekind collaborated with Weber on editing Riemann's collected works. This was likely the inspiration for their groundbreaking joint paper of 1882, "Theory of algebraic functions of a single variable," in which they put part of Riemann's work on abelian functions, which depended on the unproved Dirichlet Principle, into rigorous algebraic language. In particular they defined the Riemann surface of an algebraic curve as the prime ideals of a certain ring, and gave an algebraic proof of the important Riemann–Roch theorem. The fundamental idea of their approach was to carry over to algebraic function fields the ideas which Dedekind had earlier introduced for algebraic number fields. See [12], [14], and Chapter 3.2.2.

Beyond their technical achievement in putting major parts of Riemann's algebraic function theory on solid ground, their conceptual breakthrough lay in pointing to the strong analogy between algebraic number theory and algebraic geometry. This analogy proved extremely fruitful for both theories.

(b) Galois theory
As we mentioned, Dedekind lectured on Galois theory at Göttingen as early as 1856, only ten years after the publication of Galois' work and decades before it was assimilated by the broader mathematical community (cf. Chapter 2). His lectures were not published till much later, and so had no influence on the development of the subject.

But they give a clear indication of Dedekind's abstract thinking about the basic ideas of groups. The following remarkable quotation will suffice. It follows his proof of two theorems, namely that the product of "substitutions" (permutations) is associative and that they satisfy the cancellation law (hence form a group):

> The following investigations are exclusively based on the two fundamental theorems which we have proved [above], and on the fact that the number of substitutions is finite, therefore their results will be equally valid for *any domain* [set] of a finite number of *elements, things, concepts* $\varphi, \varphi^1, \varphi^{11}, \dots,$ which from φ, φ^1 admit a composition $\varphi\varphi^1$, defined arbitrarily but in such a way that $\varphi\varphi^1$ is again a member of that domain, and that this kind of composition obeys the laws expressed in both fundamental theorems. In many parts of mathematics, but especially in number theory and in algebra, we are continuously finding examples of this theory; the same methods of proof are valid here as there [13].

(c) Lattices

In two papers in 1897 and 1900 Dedekind introduced the notion of a lattice (he called it *Dualgruppe*). The motivation came from number theory, in particular properties possessed by various operations on ideals and modules (sums, products). The definition of a lattice was axiomatic [1]:

> If two operations $\pm$ on two arbitrary elements A, B of a (finite or infinite) system [set] G generate two elements $A \pm B$ of the same system G that satisfy the conditions (1), (2), (3) [below], then, regardless of the nature of these elements, G is called a dual group [lattice] with respect to the operations $\pm$.
> (1) $A + B = B + A \qquad A - B = B - A$
> (2) $(A + B) + C = A + (B + C) \qquad (A - B) - C = A - (B - C)$
> (3) $A + (A - B) = A \qquad A - (A + B) = A$

He derived various results from these identities, including the idempotent laws $A + A = A$ and $A - A = A$. His work on lattices inspired Ore and Birkhoff when they founded lattice theory as an independent subject in the 1930s [5].

(d) Linear algebra

In connection with his work in algebraic number theory Dedekind introduced many important ideas in linear algebra. The following quotation from his 1871 Supplement X to Dirichlet's *Vorlesungen* will suffice to give an indication of his clear grasp of the *abstract* ideas of linear algebra at this early stage in the development of its central tenets (see Chapter 5):

> If one calls m determinate numbers $a_1, a_2, \dots, a_m$ *dependent* or *independent of each other* [with respect to a field F containing the rationals Q and which has only a finite number of subfields] according as the equation $x_1a_1 + x_2a_2 + \dots + x_ma_m = 0$ is soluble or not in rational numbers $x_1, x_2, \dots, x_m$ that do not all vanish, then one finds by very easy considerations that we shall not

explore here that from a field F of the given sort only a *finite* number n of independent numbers $w_1, w_2, \ldots, w_n$ can be selected, and therefore that every number w of the field can always be uniquely represented in the form $w = h_1 w_1 + h_2 w_2 + \cdots + h_n w_n (1)$, where $h_1, h_2, \ldots, h_n$ designate *rational* numbers. We shall call the number n the *degree* [dimension of F over Q]; the complex [set] of n independent numbers w_i a basis for the field F [over Q], and the numbers h_i the *coordinates of the number* w with respect to this basis. Clearly every n numbers of the form (1) are also such a basis if the determinant formed from the corresponding n^2 coordinates is different from zero. To such a *transformation* of the basis by a linear substitution there corresponds a transformation of the coordinates by the so-called *transposed* substitution [12].

In a paper entitled "Analytic investigations related to Bernhard Riemann's paper on the hypotheses which lie at the foundations of geometry," which came to light only in 1966, Dedekind made evident his contribution to advanced linear algebra. According to Laugwitz, "he develops a substantial chunk of multilinear algebra in n dimensions, including the 'Gram' determinant and its relation to linear dependence and alternating forms" [14].

(e) The zeta function

The zeta function $\zeta(s)$ and its extensions and generalizations have been most important tools in analytic number theory since Riemann introduced $\zeta(s)$ in 1859. Among Dedekind's significant contributions to mathematics was his generalization, in Supplement XI of 1879, of Riemann's zeta function to algebraic number fields. If K is such a field, he defined the zeta function of K to be $\zeta_K(s) = \Sigma_I 1/N(I)^s$, where s is a real number greater than 1, the summation is over all ideals I of the ring of integers of K, and $N(I)$ denotes the norm of I. Just as Riemann's zeta function turned out to be important in the study of integer primes, so Dedekind's was instrumental in the study of the primes in the ring of integers of the algebraic number field K. Dedekind thereby found a formula giving the number of classes $h(K)$ of K, the so-called "class number" of K (it is finite for all K) [9].

8.2.5 Conclusion

It is time to sum up our account of Dedekind. In his Supplements to Dirichlet's *Zahlentheorie* he founded algebraic number theory and brought about a turning point in the evolution of abstract algebra, and in his works on the real and natural numbers he tamed the continuous by reducing it to the discrete (the arithmetization of analysis). But beyond the fundamental concepts he introduced and the important results he proved, were the methods he inaugurated. He was guided by philosophical principles in introducing many of his important innovations. "He does seem to be a great and true philosopher of the subject—a genuine philosopher, of and in mathematics," notes philosopher Stein [15]. One of his philosophical principles was a focus on intrinsic, conceptual properties over formulas, calculations, or concrete representations.

Another was the acceptance of nonconstructive definitions and proofs as legitimate mathematical methods—an attitude rare at the time.

His two very significant methodological innovations were the use of the axiomatic method outside of geometry and the institution of set-theoretic modes of thinking. The axiomatic method was just beginning to surface after 2000 years of dormancy. Dedekind was instrumental in pointing to its mathematical power and pedagogical value. His use of set-theoretic formulations, including that of the completed infinite— taboo at the time—preceded by about ten years Cantor's seminal work on the subject. Historian Edwards refers to his Supplement X (1871) as "the 'birthplace' of the modern set-theoretic approach to the foundations of mathematics" [11].

Not everyone was pleased with Dedekind's way of doing mathematics. Even among his mathematical soulmates there was discomfort. When Weber wrote to Frobenius in 1893 about the forthcoming publication of his *Lehrbuch der Algebra*, the latter responded as follows:

> Your announcement of a work on algebra makes me very happy. . . . Hopefully you will follow Dedekind's way, yet avoid the highly abstract approach that he so eagerly pursues now. . . . It is indeed unnecessary to push abstraction so far. I am therefore satisfied that you write the *Algebra* and not our venerable friend and master, who had also once considered that plan [5].

But of course Dedekind's "highly abstract approach" became commonplace in the early twentieth century. Of the early converts were Hilbert, Steinitz (Chapter 4), and Emmy Noether (Chapter 6). The latter, who edited Dedekind's works, used to say modestly that all she had done could already be found in his researches. Dedekind himself was modest and retiring and did not seek honors. But they came his way. He was elected to the Göttingen, Berlin, Rome, and Paris Academies, and received numerous other scientific honors on the occasion of the fiftieth anniversary of his doctorate.

The mathematician and historian of mathematics Harold Edwards, who was a great admirer of Kronecker's approach to mathematics, which was antithetical to Dedekind's, paid him a singular honor:

> Dedekind's legacy . . . consisted not only of important theorems, examples, and concepts, but of a whole *style* of doing mathematics that has been an inspiration to each succeeding generation [11].

References

1. I. G. Bashmakova and A. N. Rudakov, Algebra and algebraic number theory, in *Mathematics of the 19th Century*, ed. by A. N. Kolmogorov and A. P. Yushkevich, Birkhäuser, 2001, pp. 35–135. (Translated from the Russian by A. Shenitzer, H. Grant, and O. B. Sheinin.)
2. E. T. Bell, *Men of Mathematics*, Simon and Schuster, 1937.
3. K. S. Biermann, Dedekind, Richard, in *Dictionary of Scientific Biography*, ed. by C. C. Gillispie, Charles Scribner's Sons, 1981, vol. 4, pp. 1–5.

4. U. Bottazzini, *The Higher Calculus: A History of Real and Complex Analysis from Euler to Weierstrass*, Springer-Verlag, 1986.
5. L. Corry, *Modern Algebra and the Rise of Mathematical Structures*, Birkhäuser, 1996.
6. R. Dedekind, *Theory of Algebraic Integers*, translated and introduced by John Stillwell, Cambridge University Press, 1996.
7. R. Dedekind, Letter to Keferstein, in *From Frege to Gödel: A Source Book in Mathematical Logic*, 1879–1931, Harvard University Press, 1977, pp. 98–103.
8. R. Dedekind, Essays on the Theory of Numbers, Dover, 1963. (The book consists of the two essays: *Continuity and Irrational Numbers* and *The Nature and Meaning of Numbers*.)
9. J. Dieudonné (ed.), Abrégé d'histoire des mathématiques, 1700–1900, vol. 1, Hermann, 1978.
10. H. M. Edwards, Mathematical ideas, ideals, and ideology, *Mathematical Intelligencer* 1992, **14**(2): 6–19.
11. H. M. Edwards, Dedekind's invention of ideals, *Bulletin of the London Mathematical Society* 1983, **15**: 8–17.
12. W. Ewald (ed.), *From Kant to Hilbert: A Source Book in the Foundations of Mathematics*, Oxford University Press, 1996, vol. II, pp. 753–837.
13. J. Ferreiros, Traditional logic and early history of sets, *Archive for History of the Exact Sciences* 1996/97, **50**: 5–71.
14. D. Laugwitz, *Bernhard Riemann*, 1826–1866, Birkhäuser, 1999. (Translated from the German by A. Shenitzer.)
15. H. Stein, Logos, logic and *Logistiké*, in *History and Philosophy of Modern Mahematics*, ed. by W. Aspray and P. Kitcher, The University of Minnesota Press, 1988.

8.3 Evariste Galois (1811–1832)

Galois was a tragic and romantic figure who died in a duel at age twenty. He was also one of the foremost mathematicians of all time. In his very brief life he created one of the great edifices of mathematics—Galois Theory—of fundamental importance to this day.

He was born October 25, 1811, in Bourg-la-Reine, a village near Paris. His father was a progressive thinker who headed the liberal party in the town. He was elected mayor in 1815, the year in which Napoleon returned from exile on the island of Elba and took control for the period known as the Hundred Days. (Later in the year Napolean was exiled by the British to St. Helena.) This was the year the monarchy was restored, and, unlike in the eighteenth century, came to accept a Charter confirming most of the gains of the French Revolution.

Galois' mother came from a family of jurists and received an education in the classics. She was his sole teacher for the first 12 years of his life, stressing the study of Greek and Latin. There is no evidence she taught him any mathematics beyond rudimentary arithmetic. But these were happy years for Galois, with no hint of the troubled times to come.

His formal education began in 1823, when he enrolled in the Collège Royal de Louis-le-Grand, a Paris preparatory school (the alma mater of Robespierre and Hugo).

Evariste Galois (1811–1832)

It was at this time that he acquired his political consciousness. During the first term the students rebelled and refused to take part in the required religious observances. Scores were expelled for their disobedience. Galois was not among them, but the severity and apparent arbitrariness of the action made a deep impression on him.

Galois' first two years at Louis-le-Grand were academically successful. He won several prizes in Greek and Latin and a number of honorable mentions. During his third year his work in rhetoric was poor and he had to repeat the year. Following this reversal he enrolled, at age fifteen, in his first mathematics course. This awakened his mathematical talent. The standard mathematics texts were no challenge. He soon came across Legendre's *Elements of Geometry* and Lagrange's "The resolution of algebraic equations" and *Theory of Analytic Functions*. These fired the young Galois' imagination. Undoubtedly he was influenced in his subsequent work on Galois Theory by Lagrange's important paper "The resolution of algebraic equations" (see Chapter 2.1). He later also read Abel's work on the subject. After these encounters with masterful mathematical works he seems to have lost all interest in his normal classes at the school.

There soon followed a series of events which proved to be traumatic for the young Galois and soured him on authority. In 1828, at age sixteen, he applied, a year earlier than normal, to the very prestigious Ecole Polytechnique. But he failed the competitive entry exams. He blamed the failure on the ignorance of the examiners, but the most likely reason was his lack of preparation and communication skills. That year his father, whom he loved dearly, committed suicide as a result of persecution by authorities for his liberal views.

8.3.1 Mathematics

Galois began to make fundamental breakthroughs in the study of solvability of equations, and in early 1829 submitted a paper on the topic to the French Academy of Sciences. The referee was Cauchy, who did not present the paper to the Academy. Many sources have claimed that he lost the paper, but recent research has found otherwise. Here is Cauchy:

> I was supposed to present to the Academy . . . a report on the work of the young Galois . . . . Am indisposed at home. I regret not to be able to attend today's session and I would like you to schedule me for the following session [5].

But at the following session Cauchy still did not present Galois' work. It appears that he was very impressed with the young mathematician's paper, for he suggested that it be revised and expanded and then submitted to the Academy for the Grand Prize in mathematics. Galois sent a revised memoir to the Academy in early 1830. This was to be judged by Fourier (and others), but he died soon thereafter. The manuscript was lost. The incident increased Galois' sense of injustice and persecution by officialdom.

Despite the various setbacks Galois did publish several papers. His first (in 1829, at age 17), "Proof of a theorem on continued periodic fractions," gave little indication of his genius, but was nevertheless published in Gergonne's celebrated *Annales de mathématiques pures et appliquées*. In 1830 he published three papers: "An analysis of a memoir on the algebraic resolution of equations," "Notes on the resolution of numerical equations," and "On the theory of numbers." The first two gave indications of his evolving thoughts on what is now known as Galois theory. The third was an important paper dealing with finite fields (see Chapter 4.4).

In January 1831, at the invitation of Poisson, Galois submitted a third—his most mature—version of the memoir on the solvability of equations (recall the fate of the other two, submitted to Cauchy and Fourier, respectively). Poisson recommended that the Academy reject the paper, encouraging Galois to expand and clarify the exposition. Little wonder: the ideas were pioneering and the writing was terse. Here is an excerpt of Poisson's report:

> We have made every effort to understand Galois' proof. His reasoning is not sufficiently clear, sufficiently developed, for us to judge its correctness, and we can give no idea of it in this report. The author announces that the proposition which is the special object of his memoir is part of a general theory susceptible of many applications. Perhaps it will transpire that the different parts of a theory are mutually clarifying, are easier to grasp together than in isolation. We would then suggest that the author should publish the whole of his work in order to form a definite opinion. But in the state which the part he has submitted to the Academy now is, we cannot propose to give it approval [7].

8.3.2 Politics

We must backtrack somewhat and recount Galois' political activism—a crucial part of his story. But first some background.

The major problem facing the French authorities in the period 1815–1830 was how to bring about a compromise between the desire of those citizens who saw the gains of the Revolution as irreversible (those among them who wanted to abolish the monarchy were known as Republicans) and those who wanted to resurrect the old regime (the extreme among them were labelled Ultra Royalists). The resulting political tensions turned at times into turmoil.

The reigning King of France since 1824 was Charles X, and his cabinet comprised the Ultras. What France needed, in his view, was a return to the principle of divine right and a restoration of the authority of the Catholic Church. Elections in 1827 brought great gains for liberals and moderates. This encouraged the left to turn hostile towards the regime. In the elections of March 1830 the Republicans prevailed. The King's response was to dissolve the newly elected Chamber and to curtail the freedom of the press. Confrontation appeared inevitable. The public revolted, the insurrection lasting three days, July 27–29. On July 30 Charles X was forced to abdicate, and a compromise was reached to install the moderate Louis-Philippe as King.

Back to Galois. Having failed a second time the entrance exams for the Polytechnique, he passed the exams for the less prestigious École Normale, and began his studies in early 1830. During the "Three Glorious Days" of the revolution Galois and his fellow students were locked in by the director of the school. Galois was indignant and published a fierce attack on him in the press. He was promptly expelled.

He soon joined the National Guard, a branch of the militia composed almost entirely of Republicans. He became more and more agitated and radicalized. Sophie Germain, in a letter to her colleague Libri in early 1831, gives an indication of Galois' behavior and state of mind:

> Your preoccupation, that of Cauchy, the death of M. Fourier, have been the final blow for this student Galois who, in spite of his impertinence, showed signs of a clever disposition. ... He is without money and his mother has very little also. Having returned home [after his expulsion], he continued his habit of insult, a sample of which he gave you after your best lecture at the Academy. The poor woman fled her house, leaving just enough for her son to live on. ... They say he will go mad, I fear this is true [5].

In May 1831 a banquet was held to celebrate the acquittal of nineteen members of the Artillery of the National Guard, charged with an attempt to overthrow the government. Galois attended the festivities. He was seen with a glass of wine in one hand and a dagger held in a threatening posture in the other. Several days later he was arrested, but was acquitted at a subsequent trial, probably because of his youth.

On July 14 1831, Bastille Day, Galois and a friend were at the head of a Republican demonstration, wearing uniforms of the disbanded Artillery Guard and carrying arms—both forbidden activities. Galois was arrested and sentenced to six months in prison. A cholera epidemic saw him transferred to a hospital as a precaution.

In early 1832 he had his one and only, brief, love affair with a Stéphanie-Felicie Poterin du Motel, the daughter of a physician at Sieur Faultier, the pension at which Galois spent the last few months of his life. She terminated the affair. Galois was devastated. On May 25, five days before his death, he wrote to his best friend, Chevalier, a reputable journalist: "How can I console myself when in one month I have exhausted the greatest source of happiness a man can have, when I have exhausted it without happiness, without hope, when I am certain it is drained for life?" [5].

8.3.3 The duel

Several days later he was challenged to a duel, in circumstances which remain unclear. Several theories have been advanced by various authors writing on the subject. For example, Bell [1] and Infeld [3] claim that the duel was instigated by the secret police, who wanted to be rid of a political troublemaker. Rothman [5], in a scathing attack on Bell's version, and, more generally, on his account of Galois' life, demolishes that claim (and several others).

Rigatelli [4] asserts that the duel was a disguised suicide on Galois' part, aided by his friends, with the aim of spreading a rumour that he was murdered by the police, thus hoping to incite an insurrection. Cooke [2] casts doubt on this version.

Another account claims that the duel was instigated by his former love, Stéphanie, who had a political, or possibly a romantic, motive (see [5], [7] for details). There is some evidence for this in a sentence in Galois' letter of May 25 to Chevalier: "I die the victim of an infamous coquette and her two dupes" [5].

The known facts, according to Rothman [5], give a more nuanced, though not definitive, picture. Here is Stewart [7], quoting Rothman (probably from the Internet, no longer available) concerning the reason for the duel: "We arrive at a very consistent and believable picture of two old friends falling in love with the same girl and deciding the outcome by a gruesome version of Russian roulette" [7]. Galois' adversary in the duel was Pescheux D'Herbinville, a good friend and fellow Republican. The "Russian roulette" refers to the likelihood that only one of the two pistols (the weapons with which the duel was fought) was loaded. Several of the above points are supported by an article in the paper *Le Precursor*, written on June 4, 1832, four days following Galois' death:

> A deplorable duel ... has deprived the exact sciences of a young man who gave the highest expectations, but whose celebrated precocity was lately overshadowed by his political activities. The young Evariste Galois ... was fighting with one of his old friends, a young man like himself, like himself a member of the Society of Friends of the People, and who was known to have figured equally in a political trial. It is said that love was the cause of the combat. The pistol was the chosen weapon of the adversaries, but because of their old friendship they could not bear to look at one another and left the decision to blind fate. At point-blank range they were each armed with a pistol and fired. Only one pistol was charged [7].

Galois died on May 31, the day following the duel. Over two thousand Republicans were present at the funeral.

8.3.4 Testament

On the night before the duel, May 29, 1832, believing that he would die the next day, Galois wrote a long letter—his mathematical testament—to his friend Chevalier.

He included three manuscripts, summarized their contents, and made comments and corrections. One of the papers was the one rejected by Poisson. The following are brief excerpts from Galois' letter:

My Dear Friend,

I have made some new discoveries in analysis. The first [of the enclosed papers] concern the theory of equations, the others integral equations. In the theory of equations I have researched the conditions for the solvability of equations by radicals; this has given me the occasion to deepen this theory and describe all the transformations possible on an equation even though it is not solvable by radicals [5].

Next to one of the theorems in the Poisson paper he noted: "There are a few things to be completed in this proof. I have not the time." He concluded his letter with a request [5]:

In my life I have often dared to advance propositions about which I was not sure. But all I have written down here has been clear in my head for over a year, and it would not be in my interest to leave myself open to the suspicion that I announce theorems of which I do not have complete proof.

Make a public request of Jacobi or Gauss to give their opinions not as to the truth but as to the importance of these theorems. After that, I hope some men will find it profitable to sort out this mess.

I embrace you with effusion. E. Galois.

8.3.5 Conclusion

In his groundbreaking work on what came to be known as Galois Theory, he founded group theory, and established the field concept as fundamental in the study of solvability of equations (see Chapters 2.1 and 4.1). His entire mathematical opus comprised less than 100 pages! It is remarkable, given his age and the character of much of the mathematics of the early nineteenth century, that his focus in the study of solvability of equations by radicals was *not* on computations but on the underlying *concepts*. In his letter to Chevalier he also summarized his work in analysis—on elliptic functions and on abelian integrals, subjects which proved to be of foremost importance in the nineteenth century [8]. See Chapter 2 for details.

Galois' work on the solvability of equations by radicals was slow to be accepted. It came to the mathematical public's attention only in 1846, when Liouville published it in the prestigious *Journal de Mathématiques pures et appliqués*. But it became an integral part of the body of mathematics in 1870, with the publication of Jordan's *Traité des substitutions* (see Chapter 2.2.1). Historian Taton comments:

Galois' terse style, combined with the great originality of his thought and the modernity of his conceptions, contributed as much as the delay in publication to the length of time that passed before Galois' work was understood, recognized at its true worth, and fully developed [8].

We conclude with an appreciation of Galois by Rothman:

> There exist many fragments which indicate Galois pursued his mathematical researches, not only while in prison, but right up until the time of his death. The fact that he could work through such a turbulent life is testimony to the extraordinary fertility of his imagination. There is no question that Galois was a great mathematician who developed one of the most original ideas in the history of mathematics [5].

References

1. E. T. Bell, *Men of Mathematics*, Simon and Schuster, 1937.
2. R. Cooke, Review of *Evariste Galois, 1811–1832*, by L. T. Rigatelli (see reference 4 below), *Amer. Math. Monthly* 1998, **105**: 284–288.
3. L. Infeld, *Whom the Gods Love: The Story of Evariste Galois*, Whittlesey House, 1948.
4. L. T. Rigatelli, *Evariste Galois, 1811–1832*, Birkhäuser Verlag, 1996.
5. T. Rothman, Genius and biographers: the fictionalization of Evariste Galois, *Amer. Math. Monthly* 1982, **89**: 84–106.
6. T. Rothman, The short life of Evariste Galois, *Scient. Amer.* 1982, **246**(4): 136–149.
7. I. Stewart, *Galois Theory*, 3rd ed., Chapman & Hall, 2004.
8. R. Taton, Galois, Evariste, in *Dictionary of Scientific Biography*, ed. by C. C. Gillispie, Charles Scribner's Sons, 1981, vol. 5, pp. 259–265.

8.4 Carl Friedrich Gauss (1777–1855)

Gauss was born in Brunswick, Germany, the only son of working class parents. His father was "worthy of esteem [but] domineering, uncouth and unrefined," according to Gauss [7]. His mother was intelligent and of strong character, but only semiliterate. Gauss was a most precocious child and joked later in life that he could count before he could talk. At age eight he astonished his teacher by finding, almost instantly, the sum of the first hundred integers. The Duke of Brunswick, who had heard of his reputation, became his patron when he was fourteen and supported his education for about ten years.

In 1792 he entered the Collegium Carolinum, studying classical languages and, on his own, the works of Newton, Euler, and Lagrange. In 1795, when he enrolled at the University of Göttingen, he was still undecided about which of his two intellectual loves—philology and mathematics—he would pursue as a career.

He opted for mathematics the following year, when he proved that the regular polygon of seventeen sides is constructible with straightedge and compass. This was not just a personal triumph; it was the first discovery of a constructible regular polygon in over 2000 years (the ancient Greeks knew how to construct regular polygons of 3, 4, 5, and 15 sides). Another early landmark was Gauss' proof in 1799 of the Fundamental Theorem of Algebra, which eluded d'Alembert, Euler, and Lagrange (see Chapter 1.7). He considered the theorem so important that he gave four proofs of

Carl Friedrich Gauss (1777–1855)

it during his lifetime. The one in 1799 earned him a Ph.D. degree from the University of Helmstedt.

Gauss married happily in 1805. He remarried, unhappily, a year after the death of his wife in 1809, from which he never fully recovered. He had three children with each of his wives. He achieved a peaceful home life only in 1831, following the death of his second wife, at which time his younger daughter took over the household duties and "became the intimate companion of his last twenty-four years" [7].

Gauss made groundbreaking contributions in all areas of mathematics to which he turned: algebra, analysis (both real and complex), geometry (differential and noneuclidean), number theory, probability, and statistics. He was the Prince of Mathematicians to his contemporaries and is, by universal acknowledgment, one of the three greatest mathematicians of all time (the other two are Archimedes and Newton).

8.4.1 Number theory

Number theory, the Queen of Mathematics according to Gauss, was his first and greatest mathematical love. The *Disquisitiones Arithmeticae*, arguably his greatest work, was completed in 1798, when he was twenty-one (!), but was not published till 1801 [4]. In the seventeenth and eighteenth centuries number theory consisted of a collection of isolated, though brilliant, results. In the *Disquisitiones* Gauss systematized the subject, solved a number of its difficult and central problems, and pointed directions for future researchers.

The *Disquisitiones* began with the definition of congruence—another first. This offered an excellent example of the power of a felicitous notation, here used for the familiar idea of divisibility. A major achievement was the proof of one of the central theorems of number theory, the *quadratic reciprocity law*, already conjectured by Euler and Legendre and rediscovered by Gauss at age seventeen. It describes the relationship between the solvability of $x^2 \equiv p \pmod{q}$ and $x^2 \equiv q \pmod{p}$ for odd primes p and q (see Chapter 3.2.1). "This theorem has inspired some deep ideas of modern algebra and is of great importance throughout number theory and in other branches of mathematics" [8]. Gauss called it "the golden theorem" (theorema aureum) and during his lifetime proved it in eight different ways.

Another fundamental accomplishment in the *Disquisitiones* is the beautiful and intricate theory of *binary quadratic forms*, which investigates the representation of integers by forms $f(x, y) = ax^2 + bxy + cy^2$. (Fermat's problem on the representation of integers as sums of two squares, $n = x^2 + y^2$, is a very special case.) Gauss defined a difficult and subtle composition on such forms. He also defined an equivalence relation on the forms and showed—without using the terminology—that the equivalence classes of binary quadratic forms with a given discriminant are an abelian group under composition. This result inspired Dirichlet, Kummer, and Dedekind, among others, to try to gain *conceptual* understanding of composition of forms as defined by Gauss. Dedekind succeeded by means of his theory of ideals. See Chapter 8.2.1; also Chapters 2.2 and 3.2.

The final section of the *Disquisitiones*, a beautiful blend of algebra, geometry, and number theory, deals with *cyclotomy*—the division of a circle into n equal parts. Algebraically, it asks for the solution of $x^n - 1 = 0$. Gauss showed that this so-called cyclotomic equation is solvable by radicals for every positive integer n. This was an important result in the program, initiated by Lagrange in 1770 and brought to fruition by Galois about 1830, of determining which polynomial equations are solvable by radicals. See Chapter 2.1 and 2.2.

An important by-product of Gauss' results on cyclotomy was the characterization of regular polygons constructible with straightedge and compass: a regular n-gon is constructible if and only if $n = 2^k p_1 p_2 \ldots p_s$, where the p_i are distinct primes of the form $2^{2^t} + 1$, the so-called Fermat primes. Gauss proved the sufficiency of the condition for constructibility (the harder part) and merely asserted its necessity; this was shown in 1837 by Wantzel.

Little wonder that the *Disquisitiones* made Gauss an instant celebrity. Its wealth and the profundity of its ideas are still being mined. See [2], [3], [6], [8] for further details.

Gauss returned to number theory in 1831, introducing another groundbreaking idea. This appeared in a paper on *biquadratic reciprocity*, which investigates the relation between the solvability of $x^4 \equiv p \pmod{q}$ and $x^4 \equiv q \pmod{p}$. He found that just to *state* the law of biquadratic reciprocity he needed to introduce what came to be known as the *Gaussian integers*, defined as $G = \{a + bi : a, b \varepsilon Z\}$. He carefully analyzed the arithmetical structure of G, showing that its elements can be written uniquely as products of "primes" (in G), that is, that G is a unique factorization domain. Here was an important contribution to the founding of a new subject—*algebraic number theory* (see Chapter 3.2). This was also the paper in which Gauss defined the complex numbers as points in the plane. Although complex numbers had been used for a century, it was Gauss' sanction that made them at long last respectable as bona fide mathematical entities.

8.4.2 Differential Geometry, Probability, and Statistics

Students of astronomy and physics salute Gauss as one of their own. In 1807 he accepted, for economic reasons, the directorship of the Göttingen Observatory, a position he held for the rest of his life. He thenceforth made important contributions to both the theoretical and observational aspects of astronomy and to various branches of physics, including mechanics, optics, acoustics, and geomagnetism. But he always sought the mathematical connection, and in one instance in particular his efforts bore exceptional fruit.

In 1820 Gauss was asked by the Kingdom of Hanover (to which Göttingen belonged) to supervise a geodesic survey, which lasted several years. A major task was the precise measurement of large triangles on the earth's surface. The stimulus (presumably) provided to Gauss' fertile mind gave birth in 1827 to his famous paper on curved surfaces, in which he formulated the fundamental notion of (Gaussian) *curvature* and founded the study of the *intrinsic geometry* of curved surfaces (cf. the intrinsic geometry on the surface of a sphere). Riemann built on these ideas in the 1850s to found the theory of *n*-dimensional manifolds, which later proved indispensable in Einstein's general theory of relativity [2], [3].

Related to Gauss' astronomical work, particularly his calculation of orbits of asteroids and comets, were achievements in probability and statistics. In 1809, in a paper on the "Theory of motion of heavenly bodies," he introduced the *method of least squares* (independently found by Legendre) for obtaining the "best fit" to a series of experimental observations. In this connection he devised what came to be known as *Gaussian elimination* for the solution of a system of linear equations. In the same work he also showed that the distribution of errors when using the least squares method is "normal." This is the source of the *Gaussian (normal) distribution*, represented graphically by a bell-shaped curve [2], [6].

8.4.3 The Diary

Ideas no less profound and far-reaching than those present in Gauss' *published* works (of which we have discussed only some) were found in his *mathematical diary* [5]. This is a remarkable 19-page document of 146 very brief, often cryptic, entries dealing with discoveries covering the years 1796–1814. The diary finally became public in 1898. The first entry, dated March 30, 1796, notes Gauss' discovery of the constructibility of the regular 17-gon: "The principles upon which the division of the circle depend, and geometrical divisibility of the same into seventeen parts, etc." [5].

The publication of almost any of the 146 entries would have made mere mortals famous. Some of the entries anticipated major creations of nineteenth-century mathematics: complex analysis, elliptic function theory, and noneuclidean geometry. One can speculate about why Gauss did not publish these discoveries—which, it has been suggested, would have advanced the development of mathematics by half a century. Perhaps his motto "pauca sed matura" (few but ripe) had something to do with it. For more on this issue see [2], [8].

A comment on noneuclidean geometry is perhaps in order, especially since it may shed some light on Gauss' character. There is no doubt that he was in possession of the elements of noneuclidean geometry about two decades prior to its publication in the 1830s by J. Bolyai and by Lobachevsky. When Bolyai's father, W. Bolyai, wrote to his friend Gauss about his son's discovery, Gauss responded that he could not praise the work since he had done all that years earlier. The younger Bolyai was greatly disappointed and died without proper recognition of his great achievement. However, Gauss did praise the number-theoretic work of the young Eisenstein, "who had been one of the few to tell Gauss anything he did not already know" [7]. I will leave to readers to decide to what extent Gauss was prepared to recognize mathematical talent in others. To pursue the matter see, e.g., [1], [2], [7], [8].

8.4.4 Conclusion

Gauss was a transitional figure in the evolution of mathematics, and in particular of algebra. Stewart put it well:

> In many ways Gauss stood at the crossroads. He can be viewed equally well as either the first of the modern mathematicians or the last of the great classical ones. The paradox can easily be resolved: his methods were modern in spirit but his choice of problems was classical [8].

Among his works that inspired the development of abstract algebra are those on quadratic forms, reciprocity laws, and cyclotomy. See Chapters 2, 3, and 4 for details.

The nineteenth century witnessed fundamental transformations in mathematics, among them a growing insistence on rigor. Gauss was a leading exponent of this emerging spirit, which began to permeate all areas of mathematics. For example, in his important 1812 work on the *hypergeometric series* he was the first to insist on a rigorous treatment of convergence of series [2], [3]. "It is demanded of a proof that all doubt become impossible," he wrote to a friend. And he practiced what he preached. His proofs were elegant and polished, often to the point where all traces of his method of discovery were removed. "He is like the fox, who erases his tracks in the sand with his tail," deplored Abel [8].

The finished product of Gauss' researches gives no indication of his great skill in, and love of, calculation. Some of his deepest theorems in number theory were inspired by calculation. For example, he conjectured the *Prime Number Theorem*, namely that $\pi(x) \sim x/(\log x)$, where $\pi(x)$ is the number of primes $\leq x$ and "$\sim$" denotes "asymptotic," by first putting together a table of all primes up to 3,000,000. "It was likely a striking, and possibly unique, combination of remarkable insight, formidable computing ability, and great logical power that produced a mathematician whose ideas are still bearing rich fruit today, two centuries after he burst on the mathematical scene."

References

1. E. T. Bell, *Men of Mathematics*, Simon and Schuster, 1937.
2. W. K. Bühler, *Gauss: A Biographical Study*, Springer-Verlag, 1981.
3. G. W. Dunnington, *Carl Friedrich Gauss: Titan of Science*, Hafner Publ., 1955.
4. C. F. Gauss, *Disquisitiones Arithmeticae*, translated by A. A. Clark, Springer-Verlag, 1986.
5. J. J. Gray, A commentary on Gauss' mathematical diary, 1796–1814, with an English translation, *Expositiones Mathematicae* 1984, **2**: 97–130.
6. T. Hall, *Carl Friedrich Gauss: A Biography*, The M.I.T. Press, 1970.
7. K. O. May, Gauss, Carl Friedrich, in *Dictionary of Scientific Biography*, ed. by C. C. Gillispie, Charles Scribner's Sons, 1981, vol. 5, pp. 298–315.
8. I. Stewart, Gauss, *Scientific American* July 1977, **237**: 122–131.

8.5 William Rowan Hamilton (1805–1865)

Hamilton, born in Dublin, was the greatest Irish mathematician. At age three he was sent, likely because of financial difficulties at home, to live with his uncle James Hamilton, in Trim, County Meath. James, an Anglican clergyman schooled in the classics, supervised Hamilton's pre-university education. In his seventh year the precocious child knew (besides English) nine languages, and at age thirteen he was conversant with thirteen: Greek, Latin, Hebrew, Syriac, Persian, Arabic, Sanskrit, Hindustanee, Malay, French, Italian, Spanish, and German. When a new ambassador arrived from Persia, Hamilton (then fourteen) composed a welcoming letter in Persian, which he attempted to deliver, without success, to the ambassador.

He stayed with his uncle until he was eighteen, visiting home on vacations, and writing regularly to his parents (he was orphaned at fourteen). Here is an excerpt from one such letter, composed when he was thirteen:

> I sometimes feel as if the bottle of my brain were like those mentioned, I think in Job, 'full and ready to burst'; but when I try to uncork and empty it, like a full bottle turned upside down, its contents do not run out as fluently as might be expected; nor is the liquor that comes off as clear as could be wished.
>
> Perhaps I am not long enough in bottle to be decanted. I fear the vintage of my brain is yet too crude and unripe to make good wine of. When it shall have been more matured, I hope the produce of the vineyard you have planted and watered will afford some cups 'to cheer but not to inebriate you,' at least not shame you [11].

Aside from languages, Hamilton also studied geography, religion, literature, astronomy, and mathematics. He read Euclid (in Greek), Newton (in Latin), and Laplace (in French). At seventeen he found an error in the latter's renowned *Mécanique Céleste*,

William Rowan Hamilton (1805–1865)

having to do with the parallelogram of forces. In the same year he submitted to the Irish Academy a long essay entitled "Theory of Systems of Rays" which contained the germ of his later important work in optics. The referees suggested that the paper be clarified and expanded. The revised work was published, under the same title, in 1827.

In 1823 Hamilton entered Trinity College of the University of Dublin, placing first among one hundred candidates in the entrance exams. In his second year he won an *optime* in Greek and in his third year an *optime* in mathematical physics. "The winning of even a single *optime* was very rare" [2]. "He immediately became a celebrity in the intellectual circles of Dublin" [8]. He got his B.A. in 1827, but in 1826, while still a student, was appointed Andrews Professor of Astronomy at Trinity College and Astronomer Royal of Ireland (!). He moved to Dunsink Observatory (near Dublin) the following year and remained there the rest of his life.

His personal life was not happy. He made proposals of marriage to two young ladies, in 1825 and 1831 respectively, both of which were rejected. In 1833 he married a Helen Bayly, who suffered chronic ill health, was absent for long periods of time, and was unable to look after their three children (two sons and a daughter). Hamilton became despondent and turned to alcohol.

He showed a keen interest in poetry, writing numerous poems and receiving honors for some of them. At one point he considered becoming a poet, but was discouraged by his friend Wordsworth, who put the matter diplomatically:

> You send me showers of verses which I receive with much pleasure, as we do all; yet have we fears that this employment may seduce you from the path of science which you seem destined to tread with so much honor to yourself and

profit to others. Again and again I must repeat that the composition of verse is infinitely more of an art than men are prepared to believe, and absolute success in it depends upon innumerable *minutiae* which it grieves me you should stoop to acquire a knowledge of. ... Again I do venture to submit to your consideration, whether the poetical parts of your nature would not find a field more favorable to their exercise in the regions of prose; not because those regions are humbler, but because they may be gracefully and profitably trod, with footsteps less careful and in measures less elaborate [8].

Hamilton made groundbreaking contributions in three areas: optics (1827), dynamics (1834), and algebra (complex numbers, foundations of algebra, and quaternions; 1833 and 1843). We discuss each in turn. We will focus on algebra, touching only briefly on the other two subjects.

8.5.1 Optics

For over a century the conflict between the corpuscular and wave theories of light had remained unresolved. Newton opted for the corpuscular theory, while Huygens, and later Fresnel, pointed to wave theory. Hamilton's great achievement was to find the common theoretical basis for both standpoints. His position was stated in a letter to his friend Coleridge:

My aim has been, not to discover new phenomena, nor to improve the construction of optical instruments, but with the help of the Differential or Fluxional Calculus to remold the geometry of light by establishing one uniform method for the solution of all problems in that science, deduced from the contemplation of one central or characteristic relation ... my chief desire and direct aim being to introduce harmony and unity into the contemplation of reasonings of Optics, considered as a portion of pure Science. It has not even been necessary, for the formation of my general method, that I should adopt any particular opinion respecting the nature of light [5].

Hamilton's ideas were contained in the paper "Theory of Systems of Rays" (1827). The following quotation from the abstract that he presented to the Royal Irish Academy gives a good sense of the paper's objective:

A Ray, in Optics, is to be considered as a straight or bent or curved line, along which light is propagated; and a *System of Rays* as a collection or aggregate of such lines, connected by some common bond, some similarity of origin or production, in short some optical unity. Thus the rays which diverge from a luminous point compose one optical system, and, after they have been reflected in a mirror, they compose another. To investigate the geometrical relations of the rays of a system of which we know (as in these simple cases) the optical origin and history, to inquire how they are disposed among themselves, how they diverge or converge, or are parallel, what surfaces or curves they touch or cut, and at what angles of section, how they can be combined in partial pencils, and how each ray in particular can be determined and distinguished from every other, is to study that System of Rays. And to generalize this study of one system so as to become able to pass, without change of plan, to the study of other systems, to assign general rules and a general method whereby these separate optical arrangements may be connected and harmonized together, is to form a *Theory of Systems of Rays*. Finally, to do this in such a manner as to make available the powers of the modern mathesis, replacing figures by functions and diagrams by formulas, is to construct an Algebraic Theory of such Systems, or an *Application of Algebra to Optics* [1].

An important aspect of Hamilton's theory was its generality. His accomplishments are summarized by Hankins, who also provides technical details.

By his method it is possible to include all the properties of any optical system in a single [partial differential] equation that, when solved, will give [what Hamilton calls] a "characteristic function" completely describing the optical system. In other words, given any ray of light entering the optical system (telescope, microscope, or any system of lenses and mirrors), the characteristic function describes how the ray emerges at the other end [5].

Solving the partial differential equation in closed form was no simple matter, and one had to resort to other methods to determine properties of its solution, but on the whole Hamilton showed little interest in such "practical" matters. He put it thus:

Even if it should be thought that no practical facility is gained, yet an intellectual pleasure may result from the reduction of the most complex ... of all researches respecting the forces and motions of body, to the study of one characteristic function, the unfolding of one central relation [4].

As for practical matters, he predicted from his general studies that "a single ray incident in the correct direction on a biaxial crystal should be refracted into a cone in the crystal that emerges as a hollow cylinder" [4]. This theoretical prediction was soon verified by his colleague at Trinity, Humphrey Lloyd. It caused a sensation. "It was one of those rare events where theory predicted a completely unexpected physical phenomenon" [4].

8.5.2 Dynamics

Hamilton's work in dynamics is contained in two essays of 1834 entitled "On a General Method in Dynamics." These studies grew out of his discoveries in optics and closely paralleled them. The central connecting idea was the Principle of Least Action (already used by Lagrange in the eighteenth century in his *Mécanique Analytique*, and in its optical version, known as the Principle of Least Time, by Fermat in the seventeenth century). The crux of the matter was Hamilton's realization that a "characteristic function," which in its dynamics version he called "principal function," can be constructed whenever a minimum principle is involved. Given this version of the "characteristic function" Hamilton discovered a new form of the equations of motion, first formulated in full generality by Lagrange. Hamilton's formulation, considerably simpler than Lagrange's, was to play a central role in the early twentieth century with the rise of quantum mechanics (see below). Hankins gives a summary of the essence of Hamilton's two essays on dynamics:

> The two papers are masterpieces of analysis; there is almost no physics as such in either one.
>
> Hamilton begins the first paper by showing how the characteristic function in optics can be used analogously in dynamics. He ... shows how the characteristic function can be found from the simultaneous solution of two partial differential equations of the first order and the second degree—the same method that he had employed in the "Theory of Systems of Rays." He then shows how different forms of the equations of motion can be deduced from his characteristic function and also shows that the equations are invariant under coordinate transformation. Finally, he applies his method to problems in celestial mechanics, the subject on which he first tested his ideas.
>
> The second essay contains material more familiar to the modern student of Hamilton's dynamics. At the end of the first essay Hamilton had derived an "auxiliary" function, which, in the second essay, he named the "principal function." He then derives the "canonical equations" of motion, "Hamilton's principle" and Hamilton's version of the Hamilton–Jacobi equation. The rest of the paper is devoted to a study of perturbations of planetary orbits following the method of his dynamics [5].

Two points about the above quotation:

(a) "Hamilton's principle" says (in a modern formulation) that if the configuration of a system of moving particles is governed by their mutual gravitational attractions, then their actual paths will be minimizing curves for the integral, with respect to time, of the difference between the kinetic and potential energies of the system [10]. This is a result in the calculus of variations.

(b) Hamilton's work in dynamics was not properly appreciated by his contemporaries. One of the few exceptions was Jacobi, who developed the theory further. In particular he reduced the two partial differential equations satisfied by the principal function to a single (simpler) equation, now called the Hamilton–Jacobi equation (see [4]).

Hamilton himself valued most his work on quaternions (see below). However, in the twentieth century, especially with the rise of quantum mechanics, it was his work in dynamics that was most prized, especially the optical-mechanical analogy. This influenced such physicists as Sommerfeld, de Broglie, and especially Schrödinger, who was effusive in his praise:

> I daresay not a day passes—and seldom an hour—without somebody, somewhere on this globe, pronouncing or reading or writing Hamilton's name. That is due to his fundamental discoveries in general dynamics. The Hamilton Principle has become the cornerstone of modern physics, the thing with which a physicist expects every physical phenomenon to be in conformity....

The modern development of physics is constantly enhancing Hamilton's name. His famous analogy between optics and mechanics virtually antici- pated wave mechanics, which did not have much to add to his ideas and had only to take them more seriously [3]. See also [5].

We now turn to Hamilton's work in algebra, which was in three areas: complex numbers, the foundations of algebra, and quaternions. His work in the first two areas was published in an 1837 essay entitled "Theory of Conjugate Functions or Algebraic Couples; with a Preliminary and Elementary Essay on Algebra as the Science of Pure Time." The essay on "algebraic couples" was presented to the Irish Academy in 1833 and that on "algebra as the science of pure time" in 1835. We will consider them in the order in which they were conceived.

8.5.3 Complex Numbers

Hamilton's interest in algebra was aroused around 1826 by his mathematician friend John Graves, when Hamilton was involved in the study of optics. Graves was trying to define logarithms for negative and complex numbers (presumably unaware that Euler had done that in the previous century), and regularly corresponded with Hamilton about the problem. In 1828 he suggested that Hamilton read Warren's recently pub- lished *Treatise on the Geometrical Representation of the Square Roots of Negative Quantities*, which expressed the complex numbers as points in the plane. Hamilton was dissatisfied with a *geometric* representation of a system of *numbers* which he believed to be the domain of *algebra*. He objected in particular to the dependence of a geometric representation on a coordinate system. He was also unhappy with the representation of complex numbers as expressions of the form $a + bi$. It seemed to him that adding bi to a was like adding oranges to apples. And what, in any case, is i, he asked?

Hamilton believed in asking fundamental questions and getting to the bottom of things. The above misgivings prompted him to define complex numbers as ordered pairs of reals. (This was also done by Gauss in 1831, but it remained unpublished.) He defined the four algebraic operations on pairs, as we do today, and showed that given these operations the number-couples come close to satisfying the laws of (what we call) a field: the commutative and distributive laws are given (but not the associative law, which he introduced a decade later in his work on quaternions), as are the closure laws, additive and multiplicative inverses, and the existence of the zero element. It is interesting to note his *justification* for defining the four operations as he did:

> Proceeding to operations upon number-couples, considered in combination with each other, it is easy now to see the reasonableness of the following definitions, and even their necessity, if we would preserve, in the simplest way, the analogy of the theory of couples to the theory of singles:

$$(b_1, b_2) + (a_1, a_2) = (b_1 + a_1, b_2 + a_2);$$
$$(b_1, b_2) - (a_1, a_2) = (b_1 - a_1, b_2 - a_2);$$
$$(b_1, b_2)(a_1, a_2) = (b_1, b_2)x(a_1, a_2) = (b_1a_1 - b_2a_2, b_2a_1 + b_1a_2);$$
$$(b_1, b_2)/(a_1, a_2) = ([b_1a_1 + b_2a_2]/[a_1^2 + a_2^2], [b_2a_1 - b_1a_2]/[a_1^2 + a_2^2]).$$

Were these definitions even altogether arbitrary, they would at least not contradict each other, nor the earlier principles of Algebra, and it would be possible to draw legitimate conclusions, by rigorous mathematical reasoning, from premises thus arbitrarily assumed; but the persons who have read with attention the foregoing remarks of this theory, and have compared them with the Preliminary Essay, will see that these definitions are really *not arbitrarily chosen*, and that though others might have been assumed, no others would be equally proper [7].

These remarks suggest axiomatic-like thinking, and in that sense are well advanced for their time. Hamilton, however, did not subscribe to axiomatics, as we shall see. But why would no other definitions of the operations on pairs "be equally proper?" Toward the end of his essay on complex numbers he explains:

In the THEORY OF SINGLE NUMBERS, the symbol $\sqrt{-1}$ is *absurd*, and denotes an IMPOSSIBLE EXTRACTION, or a merely IMAGINARY NUMBER; but in the THEORY OF COUPLES, the same symbol $\sqrt{-1}$ is *significant*, and denotes a POSSIBLE EXTRACTION, or a *square-root of the couple* $(-1, 0)$. In the latter theory, therefore, though not in the former, the sign $\sqrt{-1}$ may properly be employed; and we may write, if we choose, for any couple (a_1, a_2) whatever, $(a_1, a_2) = a_1 + a_2 i$ [2].

8.5.4 Foundations of Algebra

Now we turn to the "Preliminary and Elementary Essay on Algebra as the Science of Pure Time." What was the status of algebra around the mid-1820s, when Hamilton began thinking about its foundations? The subject consisted of rules for the manipulation of algebraic expressions, especially those involving negative and complex numbers, and the solution of polynomial equations (see Chapter 1). Hamilton was to call this "practical algebra." Here is some of what he thought about it, taken from the preface to his essay:

It requires no peculiar scepticism to doubt, or even to disbelieve, the doctrine of negatives and imaginaries, when set forth (as it has commonly been) with principles like these: that a *greater magnitude may be subtracted from a less*, and that the remainder is *less than nothing*; that *two negative numbers*, or numbers denoting magnitudes each less than nothing, may be *multiplied* the one by the other, and that the product will be a *positive* number, or a number denoting a magnitude greater than nothing; and that the *square* of a number, or the product obtained by multiplying that number by itself, is therefore *always positive*, whether the number be positive or negative, yet that numbers, called imaginary, can be found or conceived or determined, and operated on by all rules of positive and negative numbers, as if they were subject to those rules, *although they have negative squares*, and must therefore be supposed to be themselves neither positive nor negative, nor yet null numbers, so that the magnitudes which they are supposed to denote can neither be greater than nothing, nor less than nothing, nor even equal to nothing. It must be hard to found a SCIENCE on such grounds as these [7].

In 1830 Peacock tried to address these issues by introducing "symbolical algebra." This gave formal expression, via the Principle of Permanence of Equivalent Forms, to the rules that the various number systems should obey (see Chapter 1.8 for details). Hamilton called this "philological algebra." He rejected this approach to algebra as well. He objected to Peacock's view of the symbols of algebra as arbitrary marks without *meaning*. He later told Peacock:

> When I first encountered [your *Treatise on Algebra*] now many years ago, and indeed for a long time afterwards, it seemed to me ... that the author designed to reduce algebra to a mere system of symbols, and *nothing more*; an affair of pothooks and hangers, of black strokes on white paper, to be made according to a fixed but arbitrary set of rules: and I refused, in my own mind, to give the high name of Science to the results of such a system [5].

To Hamilton the symbols of algebra had to stand for something "real"—not necessarily material objects but at least mental constructs. It is necessary, Hamilton claimed, to "look beyond the signs to the things signified" [5]. (Isn't this a foreshadowing of the formalist-intuitionist debate almost a century later?)

Hamilton offered a third approach—"theoretical algebra"—as the foundations of the subject:

> The Study of Algebra may be pursued in three different schools, the Practical, the Philological, or the Theoretical, according as Algebra itself is accounted an Instrument, or a Language, or a Contemplation; according as ease of operation, or symmetry of expression, or clearness of thought, (the *agere*, the *fare*, or the *sapere*,) is eminently prized and sought for. The Practical person seeks a rule which he may apply, the Philological person seeks a Formula which he may write, the Theoretical person seeks a Theorem on which he may meditate [3].

What was the "theoretical" school all about? The basic element of "contemplation," which Hamilton firmly believed would provide "clearness of thought" in algebra, was *time*. Briefly, just as, according to Kant, in his *Critique of Pure Reason*, geometry is grounded in a mental intuition of space, so algebra, Hamilton argued, should be grounded in a mental intuition of time. Kant was not the only inspiration for making time a fundamental element in Hamilton's algebra. Newton was another. Hamilton claimed that he was on firm ground in focusing on time since it is fundamental in Newton's fluxional calculus (fluxions are instantaneous rates of change with respect to time). Here are some of his thoughts on theses matters:

> The argument for the conclusion that *the notion of time may be unfolded into an independent Pure Science* [viz. Algebra], or that *a Science of Pure Time is possible*, rests chiefly on the existence of certain intuitions, connected with that notion of time, and fitted to become the sources of a Pure Science, and on the actual deduction of such a Science from those principles, which the author conceives that he has begun [2].

By providing a foundation for algebra Hamilton had in mind placing the real numbers—on which, he felt, all of algebra is based—on secure grounds. This meant to him *constructing* the real numbers in terms of the pure intuition (not a temporal intuition) of time. Fundamental was also the notion of order in time:

> The notion of ORDER IN TIME is not less but more deeply seated in the human mind, than the notion or intuition of ORDER IN SPACE; and a mathematical Science may be founded on the former, as pure and as demonstrative as the science founded on the latter. There is something mysterious and transcendent involved in the idea of Time; but there is also something definite and clear: and while the Metaphysicians meditate on the one, Mathematicians may reason from the other [3].

Here is a brief description of Hamilton's treatment of the positive and negative integers:

> He considered an "equidistant series of moments" ... $E^{11}E^1EABB^1B^{11}$ ..., each letter representing an instant or moment of time, such that the intervals of time between successive moments are all equal to one another. Some moment such as A was selected "as a standard with which all the others are to be compared" and was called the *zero moment*. One of the moments near to A (such as B) was called the *positive first*. The operator or step by which one passed from any moment to the next moment to the right was denoted by a, so that $B = a + A$, $B^1 = a + a + A = 2a + A, \ldots$. The operator by which one passed from any moment to the next moment on the left was denoted by θa. Thus $A = \theta a + B$, $E = \theta a + A, \ldots$. The series of steps from the zero-moment to another moment were now denoted by $\ldots 3\theta a, 2\theta a, 1\theta a, 0a, 1a, 2a, 3a, \ldots$ [7].

These then are the integers. To us they appear to be metaphysical objects rather than mathematical entities. To Hamilton, the notion of opposite direction in time was real, while that of negative magnitude was meaningless. This was the first instalment of his "long-aspired-to work on the union of Mathematics and Metaphysics" [5]. He continued in a similar vein with constructions of the rationals and reals (see [5], [7] for details). Needless to say, Hamilton's ideas on algebra as the science of pure time had little resonance among mathematicians. In fact, Hamilton was probably the only member of the "theoretical school" of algebra which he had founded. Nevertheless, his work was a laudable attempt—the first such—at arithmetization of mathematics.

8.5.5 Quaternions

Our last topic is Hamilton's invention (discovery?) of quaternions. These arose from his attempts, begun already in 1827, to introduce multiplication on "triplets" (as he called them). Having defined complex numbers (vectors in the plane) as ordered pairs, it was natural for him to inquire whether an algebra of triplets would be possible to represent vectors in 3-space. Since the complex numbers were fundamental in many branches of mathematics and its applications, he considered the task of finding a similar algebra of triplets to be of vital importance.

Addition and subtraction of triplets were to be defined componentwise, in the obvious way. As for multiplication, he imposed several conditions it would have to satisfy: It had to be associative, commutative, and distributive (over addition), division had to be possible, the "law of the moduli" had to hold (the modulus of the product is equal the product of the moduli, where the modulus of the triplet (a, b, c) is $a^2 + b^2 + c^2$), and finally, the product of triplets had to have geometric significance, just as the product of vectors in the plane does [2].

We have pointed out that Hamilton rejected the *formal* definition of complex numbers as vectors in the plane, but he was happy to use the vector representation as an *aid to intuition*. With that in mind he represented triplets (a, b, c) also as vectors in 3-space, in the form $a + bi + cj$, where the properties of j were to be determined. In attempts at defining the product of triplets his task reduced to determining the products ij and j^2.

Hamilton worked for fifteen years trying to find a multiplication for triplets which would satisfy the conditions stated above. As we know, he did not succeed, and turned to quadruples, (a, b, c, d), which he also denoted by $a + bi + cj + dk$. A blow-by-blow account of his attempts to define products of triplets is given in a nice article by van der Waerden [12]. Hamilton's own account of these events can be found in [3]. Below he describes his "flash of insight" on the invention of the quaternions, in 1843. The account appears in a letter of 1865 to his son Archibald:

If I may be allowed to speak of *myself* in connexion with the subject, I might do so in a way which would bring *you* in, by referring to an *anti-quaternionic* time, when you were a mere *child*, but had caught from me the conception of a Vector, as represented by a *Triplet*: and indeed I happen to be able to put the finger of memory upon the year and month—October, 1843—when having recently returned from visits to Cork and Parsonstown, connected with a Meeting of the British Association, the desire to discover the laws of the multiplication referred to regained with me a certain strength and earnestness, which had for years been dormant, but was then on the point of being gratified, and was occasionally talked of with you. Every morning in the early part of the above-cited month, on my coming down to breakfast, your (then) little brother William Edwin, and yourself, used to ask me, "Well, Papa, can you *multiply* triplets?" Whereto I was always obliged to reply, with a sad shake of the head: "No, I can only *add* and subtract them."

But on the 16th day of the same month—which happened to be a Monday, and a Council day of the Royal Irish Academy—I was walking in to attend and preside, and your mother was walking with me, along the Royal Canal, to which she had perhaps driven; and although she talked with me now and then, yet an *under-current* of thought was going on in my mind, which gave at last a *result*, whereof it is not too much to say that I felt *at once* the importance. An *electric* circuit seemed to *close*; and a spark flashed forth, the herald (as I *foresaw, immediately*) of many long years to come of definitely directed thought and work, by *myself* if spared, and at all events by *others*, if I should even be allowed to live long enough distinctly to communicate the discovery. Nor could I resist the impulse—unphilosophical as it may have been—to cut with a knife on a stone of Brougham Bridge, as we passed it, the fundamental formula with the symbols, i, j, k; namely

$$i^2 = j^2 = k^2 = ijk = -1,$$

Which contains the *Solution* of the *Problem*, but, of course, as an inscription, has long since mouldered away. A more durable notice remains, however, on the council Book of the Academy for that day (October 16th, 1843), which records the fact, that I then asked for and obtained leave to read a Paper on *Quaternions*, at the *First General Meeting* of the Session: which reading took place accordingly, on Monday the 13th of the November following [2].

Another retrospective account, this time of Hamilton's *inspiration* for the introduction of the quaternions, was given in a letter of 1855 to a Rev. Townsend:

The quaternion [was] born, as a curious offspring of a quaternion of parents, say of geometry, algebra, metaphysics, and poetry. . . . I have never been able to give a clearer statement of their nature and their aim than I have done in two lines of a sonnet addressed to Sir John Herschel:

"And how the one of Time, of Space the Three,
Might in the Chain of Symbols girdled be" [5].

The geometric motivation was the desire to extend vectors in the plane to vectors in space (an extension which the quaternions, in a sense, accomplished, with vectors in 3-space represented by "pure quaternions" $ai + bj + ck$); the algebra stemmed from a natural desire to extend number-pairs to triples, and when that failed, to quadruples; the metaphysical connection with time was a factor in all of Hamilton's works in algebra (witness his sonnet, above; see also Section 8.5.4); as for the poetry, we recall that Hamilton was an aspiring poet and no doubt thought of his work in poetic terms. He called Lagrange's *Mécanique Analytique* a "scientific poem." He would also have subscribed to Weierstrass' sentiment that "No mathematician can be a complete mathematician unless he is also something of a poet."

A major reason why Hamilton did not succeed in finding the desired multiplication of triplets is that the law of the moduli does not hold: the product of a sum of three squares is a sum of four rather than of three squares. In fact, Euler had shown a century earlier that the product of a sum of four squares is again a sum of four squares. Had Hamilton known of this identity he might have proceeded with the multiplication of quadruples without spending fifteen years on triplets. But that is hindsight. In the 1860s Weierstrass showed that no n-tuples of reals forming a field extension of the complex numbers are possible, so Hamilton's attempts to multiply triples were doomed.

The quaternions form a skew field, that is, they satisfy all the axioms of a field except for commutativity of multiplication. (Assuming the associative law, the identities $ij = k = -ji, jk = i = -kj, ki = j = -ik$ follow from those which Hamilton inscribed on the stone at the Brougham Bridge.) In fact, they form a division algebra (a skew field which is also a vector space over the reals). Their invention was a breakthrough in the evolution of algebra. It detached algebra from arithmetic: now there was a number system which satisfied all the laws of arithmetic save for commutativity of multiplication. Poincaré referred to this development as "a revolution in arithmetic quite comparable to that which Lobachevsky effected in geometry" [9]. Indeed, both events were radical departures from existing canons, and both led to fundamental developments in their respective fields.

At first, however, the quaternions received less than universal acclaim. When Hamilton communicated his invention to his friend, mathematician John Graves, the latter responded as follows:

> There is still something in the system [of quaternions] which gravels me. I have not yet any clear views as to the extent to which we are at liberty arbitrarily to create imaginaries, and to endow them with supernatural properties [2].

But most mathematicians, including Graves, quickly came around to Hamilton's point of view. The floodgates were opened and the stage was set for the exploration of diverse "number systems" with properties which differed in various ways from those of the real and complex numbers. This marked the genesis of noncommutative algebra. See Chapter 3.1 for details.

The quaternions were also important in geometry, as one of the first examples of a space of dimension greater than three (Cayley and Grassmann introduced that notion independently, about the same time). On the other hand, "Hamilton was convinced that in the quaternions he had found a natural algebra of three-dimensional space" [4]. He put it thus: "And here there dawned on me the notion that we must admit, in some sense, a *fourth dimension* of space for the purpose of calculating with triplets" [12]. A more useful system of (three-dimensional) vector analysis was introduced in the 1880s independently by Gibbs and Heaviside. Crowe argues that it was Hamilton's quaternions that were instrumental in the rise of vector analysis [2]; see also [5].

For twenty-two years following the invention of the quaternions Hamilton was preoccupied almost exclusively with their application to geometry, physics, and elsewhere. To him they were the long-sought key which would unlock the mysteries of geometry and physics. "I still must assert," he noted, "that this discovery appears to me to be as important for the middle of the nineteenth century as the discovery of fluxions [the calculus] was for the close of the seventeenth" [2]. From *our* perspective, however, the importance of the quaternions lies in *algebra*.

Hamilton wrote two books on quaternions, *Lectures on Quaternions*, and *Elements of Quaternions*, published in 1853 and 1866 (posthumously), respectively. Both were very long (over 700 pages) and difficult to read, and so exerted little influence. Among other things, Hamilton attempted to extend quaternions to n-tuples for $n > 4$:

> ... the same general *view* of algebra, as the science of pure time, admitted easily, at least in thought, of an *extension* of this whole theory, not only from couples to triples, but also from triples to *sets*, of moments, steps, and

numbers. Instead of *two* or even *three* moments, there was no difficulty in conceiving a system or *set* of *n* such moments, $A_1, A_2, \ldots, A_n, \ldots$. [They] could be *added* or *subtracted*, by adding or subtracting their *component steps* ..., and ... could be *multiplied* by a number .... But when it was required to divide one [moment] by another, ... a difficulty again arose, which I proposed still to meet on the same general plan as before ... [3].

Of course, *we* know that no division algebra of *n*-tuples is possible for $n > 4$. If associativity under multiplication is given up, one other division algebra exists, namely the octonions ($n = 8$). See Chapter 3.1 for details.

8.5.6 Conclusion

We come to the end of our account of Hamilton. He was highly esteemed in Ireland and England, many honors coming his way. In 1832, at age twenty-six, he was elected a member of the Royal Irish Academy, and he served as its president from 1837 to 1845, when he resigned to devote all his time to quaternions. In 1835 he was knighted and in 1836 the Royal Society awarded him the Royal Medal for his work in optics. At thirty-eight he received a life pension from the British government. In 1848 both the Royal Irish Academy and the Royal Society of Edinburgh awarded him medals for his work on quaternions. In 1865 (the year of his death), when the US National Academy of Sciences was founded, he was made its first foreign member, mainly for his work on quaternions, which profoundly influenced Harvard mathematician Benjamin Peirce (see Chapter 3.1). Hamilton had, however, little contact with colleagues on the continent and received no recognition from them. The one exception was Jacobi, who referred to him as "the Lagrange of your country" (by "your country" he meant the English-speaking world) [1].

We end with two tributes. The first is from Hankins, Hamilton's biographer:

> The high degree of abstraction and generality that made his papers so difficult to read also made them stand the test of time, while more specialized researches with greater immediate utility have been superseded [4].

Finally, here is an unreserved appreciation by Lanczos:

> His optical and dynamical investigations were prophetic and foreshadowed the quantum theory of our days. His quaternions foreshadowed the space-time world of relativity. The quaternion algebra was the first example of a non-commutative algebra, which released an avalanche of literature in all parts of the world. Indeed, his professional life was fruitful beyond measure [6].

References

1. E. T. Bell, *Men of Mathematics*, Simon and Schuster, 1937.
2. M. J. Crowe, *A History of Vector Analysis*, University of Notre Dame Press, 1967.

3. W. Ewald (ed.), *From Kant to Hilbert: A Source Book in the Foundations of Mathematics*, Oxford University Press, 1996, vol. 1, pp. 362–441.
4. T. L. Hankins, Hamilton, William Rowan, in *Dictionary of Scientific Biography*, ed. by C. C. Gillispie, Charles Scribner's Sons, 1981, vol. 6, pp. 85–93.
5. T. L. Hankins, *Sir William Rowan Hamilton*, The Johns Hopkins University Press, 1980.
6. C. Lanczos, William Rowan Hamilton—an appreciation, *American Scientist* 1967, **2**: 129–143.
7. C. C. MacDuffee, Algebra's debt to Hamilton, *Scripta Math.* 1944, **10**: 25–35.
8. A. MacFarlane, *Lectures on Ten British Mathematicians of the Nineteenth Century*, Wiley, 1916.
9. H. Poicaré, Review of Hilbert's Foundations of Geometry, *Bull. Amer. Math. Soc.* 1999, **37**(1): 77–78. (Reprinted.)
10. G. F. Simmons, *Differential Equations*, McGraw-Hill, 1972.
11. T. D. Spearman, William Rowan Hamilton 1805–1865, *Proc. Royal Irish Acad.* 1995, **95A**: 1–12.
12. B. L. van der Waerden, Hamilton's discovery of quaternions, *Math. Mag.* 1976, **49**: 227–234.

8.6 Emmy Noether (1882–1935)

Emmy Noether was a towering figure in the evolution of abstract algebra. In fact, she was the moving spirit behind the abstract, axiomatic approach to algebra. She also had a singular personality which attracted a group of students and collaborators who spread the gospel of abstract algebra far and wide. We will give here a sketch of her life. In Chapter 6 we discussed her work, including her intellectual debts and her legacy.

8.6.1 Early Years

Noether was born in 1882 in Erlangen, the German university town of Klein's Erlangen Program fame. The university was founded in 1743 and had among its mathematics faculty such luminaries as von Staudt, Klein, Gordan (the "king of invariants"), and Max Noether, the famous algebraic geometer and Emmy's father. Gordan was a contemporary and friend of Max Noether and a frequent visitor of the Noether's. Although Emmy showed little early interest in mathematics, the frequent mathematical conversations at the Noether household between Gordan and her father were an important part of the atmosphere in which she grew up. Gordan was later to become Noether's thesis advisor.

Noether came from an economically well-established household, and her childhood seems to have been happy. She liked dancing and took piano lessons, which she did not like. She was a friendly and likeable child. Between the ages of seven and fifteen she went to the "Municipal School for Higher Education of Daughters," where she studied English and French. There is no indication that she wanted to study mathematics or science, but these were, of course, not "feminine" subjects. In 1900, at the age of eighteen, she was certified as a teacher of both subjects in "Institutions for the Education and Instruction of Females." She chose, however, not to pursue this career and instead enrolled at the University of Erlangen.

Amalie Emmy Noether (1882–1935)

8.6.2 University Studies

Easier said than done in those days. As the famous German historian Heinrich von Treitschke put it in the early 1890s:

> Many sensible men these days are talking about surrendering our universities to the invasion of women, and thereby falsifying their entire character. This is a shameful display of moral weakness. They are only giving way to the noisy demands of the Press. The intellectual weakness of their position is unbelievable. . . . The universities are surely more than mere institutions for teaching science and scholarship. The small universities offer the students a comradeship which in the freedom of its nature is of inestimable value for the building of a young man's character [4].

In 1898, two years before Noether entered the University of Erlangen, the Senate of the University declared that the admission of women students would "overthrow all academic order" [5]. By 1900, however, the authorities relented and extended to women the *conditional* right to enroll in German universities. (Women were permitted to enroll at universities in the U.S. in 1853, in France in 1861, in England in 1878, and in Italy in 1885.)

Individual professors had the right, which they often exercised, to deny women permission to attend their lectures. This meant that Noether, one of two women among

1,000 students at the university, had to choose her subjects and instructors with some care. In fact, she at first took courses in history and modern languages, but later switched to mathematics—it is not clear exactly when and why. By 1904 she was formally able to register as a student at the University of Erlangen, now studying only mathematics. In 1908 she received her Ph.D. degree, summa cum laude, having written a thesis on invariants under Gordan. See Chapter 6.1.

Between 1908 and 1915 she worked without compensation at the University of Erlangen. "Working" meant doing research, attending meetings of the German Mathematical Society and giving presentations, and occasionally substituting at lectures for her ailing father.

8.6.3 Göttingen

Although Noether did not have a formal position during these seven years, they were, as noted, not idly spent. And they bore fruit. She had become an expert on invariant theory, to the point that in 1915 Hilbert and Klein invited her to Göttingen to help them with problems on differential invariants. These proved important in connection with their work on mathematical aspects of the general theory of relativity. (It was 1915, and Einstein had just promulgated his general theory of relativity. Both Hilbert and Klein turned their attention to it.)

Noether's move to Göttingen was of singular importance. The university was at that time considered the world center of mathematics. With Gauss, Dirichlet, and Riemann as former professors, and with the contemporary faculty including Klein, Hilbert, Landau, Minkowski, and Courant, and later Weyl, Bernays, and Neugebauer, Göttingen had become the "Mecca of Mathematics." The list of visitors reads like a "who's who" of the world of mathematics: van der Waerden from Holland, Olga Taussky and Köthe from Austria, Tagaki and Shoda from Japan, André Weil and Chevalley from France, Schmidt, Gelfond, Alexandrov, Kolmogorov, and Urysohn from the Soviet Union, Tsen from China, Kuratowski from Poland, MacLane, G. D. Birkhoff, Wiener, and Lefschetz from the United States, and Artin, Hasse, Brauer, Siegel, and von Neumann from various universities in Germany.

Noether thrived in these surroundings. The decade 1920–1930 was the decisive period of her mathematical life. This is when she made her groundbreaking contributions to algebra (see Chapter 6). She was then in her forties. "Such a late maturing is a rare phenomenon in mathematics," notes Weyl [11], mentioning Sophus Lie as another great exception to the rule. She attracted students, co-workers, and visitors who vigorously developed the subject soon to become known as "modern algebra." About 1930, the algebraists around Noether had gained the reputation as the most active group at the Mathematical Institute of Göttingen—quite an accomplishment given the presence of the likes of Hilbert, Weyl, Landau, and Courant!

Two great honours came her way in 1932. First, she was awarded, jointly with Artin, the "Ackermann–Teubner Memorial Prize" for the advancement of the mathematical sciences. Second, she gave one of the twenty-one plenary addresses at the International Congress of Mathematicians in Zurich. It was a remarkable event for a woman to be invited to give a plenary talk.

But all did not go smoothly for Noether at Göttingen. One began a university career in Germany as Privatdozent, comparable in rank to an assistant professor. This was an unpaid position which gave its holder the right to teach. The income of Privatdozenten consisted of minimal fees paid by students for attending their lectures. One would have thought that under such circumstances, and having been invited to Göttingen by the great Klein and Hilbert, Noether would have got an appointment as Privatdozent immediately upon her arrival in Göttingen.

That was not to be, however. The philologists and historians of the Philosophical Faculty, of which mathematics was part, opposed Hilbert's efforts to allow Emmy to habilitate—a necessary step in becoming Privatdozent—because she was a woman. Hilbert protested, without success, to the University Senate: "After all," he claimed, "we are a university, and not a bathing establishment" [11]. Only four years later (in 1919) was Noether allowed to habilitate and become Privatdozent. This followed the war, which brought profound political and social change, including an improvement in the legal position of women.

Three years later, in 1922, the mathematics department of Göttingen applied to the Ministry of Education to appoint Noether as professor. She was given the title "extraordinary professor without tenure" ("extraordinary" is the equivalent of an associate professor). This was merely a title, carrying no obligations and no salary. Since the high postwar inflation in Germany greatly reduced students' ability to pay their instructors, Noether was fortunate to get in the following year a "Teaching Assignment" in algebra, which provided a small remuneration. It required, however, annual confirmation by the Ministry. This is the position she remained in until she left Göttingen ten years later.

Why was there little institutional recognition of Emmy Noether's talents and accomplishments? We can only speculate, of course. But she had several marks against her: she was a woman, she was Jewish, and she had leftist political sympathies.

8.6.4 Noether as a Teacher

What kind of teacher was Noether? By standard measures, she was not a good teacher. She did not give well-organized, polished lectures. Yet, she inspired many students, through her lectures and through personal contact. Here is testimony from three distinguished mathematicians who attended her courses:

> She was concerned with concepts only, not with visualization or calculation ... This ... was probably one of the main reasons why her lectures were difficult to follow ... And yet, how profound the impact of her lecturing was! (Van der Waerden [9]).

> Professor Noether's lectures ..., are ... excellent, both in themselves and because they bear an entirely different character in their excellence. Professor Noether thinks fast and talks faster. As one listens, one must also think fast— and that is always excellent training (MacLane [6]).

> To an outsider Emmy Noether seemed to lecture poorly, in a rapid and confusing manner, but her lectures contained a tremendous force of mathematical thought and an extraordinary warmth and enthusiasm (Alexandrov [1]).

Indeed, Noether had a warm and caring personality. She was also modest, generous, frank, strong-willed, and outwardly coarse. "She was both a loyal friend and a severe critic," said van der Waerden [9], giving expression to one of these seeming contradictions. Her personal traits, combined with deep mathematical insights, attracted a core of devoted students, the so-called "Noether boys." (Among her Ph.D. students were Deuring, Fitting, Grell, Greta Hermann, Krull, Levitzki, F.K. Schmidt, Ruth Stauffer, and Witt.) They often visited her home, and they used to go on frequent walks together. The topic of conversation was almost invariably mathematics. Here is the story of one such walk.

> It was raining, and Emmy Noether's umbrella did not offer much protection since it was in poor condition. When her students suggested that she get it repaired, she replied: "Quite right, but it can't be done: when it doesn't rain, I don't think of the umbrella, and when it rains, I need it" [3].

In a more serious vein, van der Waerden relates the following:

> I wrote a paper based upon this simple idea and showed it to Emmy Noether. She at once accepted it for the Mathematische Annalen, without telling me that she had presented the same idea in a course of lectures just before I came to Göttingen. I heard it later from Grell, who had attended her course [10].

On January 31, 1933 Hitler assumed the office of Chancellor. On March 31 he announced the beginning of the Third Reich. On April 25 Noether was dismissed from her teaching position. The dismissal of Courant, Landau, and Bernays followed in short order. Courant was replaced as head of the Mathematics Institute at Göttingen by Neugebauer, who lasted one day in that position. He refused to sign the required loyalty declaration.

8.6.5 Bryn Mawr

With Weyl's assistance, Noether got a visiting position at Bryn Mawr College in Pennsylvania. The transition might have been difficult but for the warm reception she received at Bryn Mawr and the mathematical contacts she established at nearby Princeton. At Bryn Mawr she had her "Noether girls"—one doctoral and three post-doctoral students (among the latter was Olga Taussky). At Princeton she began in early 1934 to give weekly lectures on algebra. Writing to Hasse about them, she said:

> I'm beginning to realize that I must be careful; after all, they are essentially used to explicit computation and I have already driven a few of them away with my approach [3].

Among those who were not driven away were Albert, Brauer, Jacobson, Vandiver, and Zariski. In a book on Zariski, Carol Parikh pointed out that "Zariski's contact with Noether was undoubtedly the single most important aspect of that year for him" [7].

The time she spent at Bryn Mawr and Princeton was the happiest in her life, Noether told Veblen before her death. She was respected and appreciated as she had never been in her own country. But it was a brief, if happy, year and a half. On April 10, 1935 she underwent an operation for a tumor. She was recovering well when, four days later, complications brought unexpected death.

8.6.6 Conclusion

Ten days after her death Hermann Weyl delivered at Bryn Mawr a moving and eloquent eulogy. Let me conclude this account of Noether's life by quoting from it:

> It was only too easy for those who met her for the first time, or had no feeling for her creative power, to consider her queer and to make fun at her expense. She was heavy of build and loud of voice, and it was often not easy for one to get the floor in competition with her. She preached mightily, and not as the scribes. She was a rough and simple soul, but her heart was in the right place. Her frankness was never offensive in the least degree. In everyday life she was most unassuming and utterly unselfish; she had a kind and friendly nature. Nevertheless she enjoyed the recognition paid her; she could answer with a bashful smile like a young girl to whom one had whispered a compliment. No one could contend that the Graces had stood by her cradle; but if we in Göttingen often chaffingly referred to her as "der Noether" (with the masculine article), it was also done with a respectful recognition of her power as a creative thinker who seemed to have broken through the barrier of sex. She possessed a rare humor and a sense of sociability; a tea in her apartment could be most pleasurable ... She was a kind-hearted and courageous being, ready to help, and capable of the deepest loyalty and affection. And of all I have known, she was certainly one of the happiest. ...
> Two traits determined above all her nature: First, the native productive power of her mathematical genius. She was not clay, pressed by the artistic hands of God into a harmonious form, but rather a chunk of human primary rock into which he had blown his creative breath of life. Second, her heart knew no malice; she did not believe in evil—indeed it never entered her mind that it could play a role among men. This was never more forcefully apparent to me than in the last stormy summer, that of 1933, which we spent together in Göttingen. ... A time of struggle like this one ... draws people closer together; thus I have a particularly vivid recollection of these months. Emmy Noether, her courage, her frankness, her unconcern about her own fate, her conciliatory spirit, were in the midst of all the hatred and meanness, despair and sorrow surrounding us, a moral solace ... The memory of her work in science and of her personality among her fellows will not soon pass away. She was a great mathematician, the greatest, I firmly believe, that her sex has ever produced, and a great woman [11].

References

1. P. S. Alexandrov, In memory of Emmy Noether, in *Emmy Noether, 1882–1935*, ed. by A. Dick, Birkhäuser, 1981, pp. 153–179.
2. J. W. Brewer and M. K. Smith (eds), *Emmy Noether: A Tribute to her Life and Work*, Marcel Dekker, 1981.
3. A. Dick, *Emmy Noether, 1882–1935*, Birkhäuser, 1981.

4. R. J. Evans, *The Feminist Movement in Germany, 1894–1933*, Sage Publ., 1976.
5. C. H. Kimberling, Emmy Noether and her influence, in *Emmy Noether: A Tribute to her Life and Work*, ed. by J. W. Brewer and M. K. Smith, Marcel Dekker, 1981, pp. 3–61.
6. S. MacLane, Mathematics at the University of Göttingen (1931–1933), in *Emmy Noether: A Tribute to her Life and Work*, ed. by J. W. Brewer and M. K. Smith, Marcel Dekker, 1981, pp. 65–78.
7. C. Parikh, *The Unreal Life of Oscar Zariski*, Academic Press, 1991.
8. B. Srinivasan and J. Sally (eds), *Emmy Noether in Bryn Mawr*, Springer-Verlag, 1983.
9. B. L. van der Waerden, Obituary of Emmy Noether, in *Emmy Noether, 1882–1935*, ed. by A. Dick, Birkhäuser, 1981, pp. 100–111.
10. B. L. van der Waerden, The foundations of algebraic geometry from Severi to André Weil, *Arch. Hist. Ex. Sc.* 1970–71, **7**: 171–180.
11. H. Weyl, Memorial address, in *Emmy Noether, 1882–1935*, ed. by A. Dick, Birkhäuser, 1981, pp. 112–152.

Index